Dredging in Coastal Waters

BALKEMA – Proceedings and Monographs
in Engineering, Water and Earth Sciences

Dredging in Coastal Waters

D. Eisma

*Department of Marine Sedimentology,
Utrecht University, The Netherlands
(Retired)*

LONDON/LEIDEN/NEW YORK/PHILADELPHIA/SINGAPORE

A Catalogue record for the book is available from the Library of Congress

Copyright © 2006 Taylor & Francis plc., London, UK

All rights reserved. No part of this publication or the information contained herein may be reproduced, stored in a retrieval system, or transmitted in any form or by any means, electronic, mechanical, by photocopying, recording or otherwise, without written prior permission from the publishers.

Although all care is taken to ensure the integrity and quality of this publication and the information herein, no responsibility is assumed by the publishers nor the author for any damage to property or persons as a result of operation or use of this publication and/or the information contained herein.

Published by: Taylor & Francis/Balkema
P.O. Box 447, 2300 AK Leiden, The Netherlands
e-mail: Pub.NL@tandf.co.uk
www.balkema.nl, www.tandf.co.uk, www.crcpress.com

ISBN 0 415 39111 3

Printed in Great Britain

Contents

Introduction   xiii

CHAPTER 1   DREDGING TECHNIQUES; ADAPTATIONS TO REDUCE ENVIRONMENTAL IMPACT   1

1.1   Introduction   1
1.2   Types of dredging projects   2
     1.2.1   Capital dredging works   2
     1.2.2   Maintenance dredging works   3
     1.2.3   Remedial dredging works   3
1.3   Criteria to judge the environmental effects of a dredger   4
1.4   Phases of a dredging project   5
     1.4.1   Dislodging of the material   5
     1.4.2   Raising the material   5
     1.4.3   Horizontal transport of the material   6
     1.4.4   Placement of the material   6
1.5   The dredging equipment   7
     1.5.1   Hydraulic dredgers   7
         1.5.1.1   Suction Dredger (SD)   7
         1.5.1.2   Cutter Suction Dredger (CSD)   9
         1.5.1.3   Trailing Suction Hopper Dredger (TSHD)   10
     1.5.2   Mechanical dredgers   12
         1.5.2.1   Bucket Ladder Dredger (BLD)   12
         1.5.2.2   Backhoe Dredgers (BHD)   13
         1.5.2.3   Grab Dredger (GD)   14
     1.5.3   Hydrodynamic dredgers   15
         1.5.3.1   Water Injection Dredger (WID)   16
         1.5.3.2   Underwater plough (UWP)   17
1.6   Recent developments in low-impact dredging equipment   18
     1.6.1   New developments in existing dredging equipment   19
         1.6.1.1   Degassing system, for hydraulic dredgers   19
         1.6.1.2   Overflow reduction, for TSHD   19
         1.6.1.3   Turtle deflecting device, for TSHD   19
         1.6.1.4   Dredging Information Systems, for all types of dredgers   19
         1.6.1.5   Monitoring and control systems for mechanical dredgers   20
         1.6.1.6   Positioning systems for mechanical dredgers   20
         1.6.1.7   Special buckets for BHD and GD   21
         1.6.1.8   Encapsulated bucket line for BLD   21
     1.6.2   Development of new dredging equipment   21
         1.6.2.1   Disc bottom cutter dredger   22

		1.6.2.2	Sweep dredger	23
		1.6.2.3	Environmental auger dredger	24
		1.6.2.4	Environmental grab dredger	26
		1.6.2.5	Pneumatic dredgers	28
1.7	Transport and disposal equipment			29
	1.7.1	Pipeline transport		29
	1.7.2	Barge transport		30
	1.7.3	Road transport		31
	1.7.4	Conveyor belt transport		32
	1.7.5	Combined transport cycles		32
1.8	Placement techniques			33
	1.8.1	On-land placement		33
	1.8.2	Underwater placement		34
	1.8.3	Capping techniques		35
1.9	Mitigating measures			36
	1.9.1	Measures at the dredging site		36
	1.9.2	Measures at the relocation site		37
1.10	Monitoring and control of the dredging process			38
	1.10.1	Objectives of the monitoring activities		38
		1.10.1.1	Ensuring compliance with restrictions	38
		1.10.1.2	Verifying project conditions	38
		1.10.1.3	Providing feedback	38
		1.10.1.4	Increasing know-how	39
	1.10.2	Planning a monitoring programme		39
		1.10.2.1	Define the critical activities	39
		1.10.2.2	Take an overall view instead of point measurements	39
		1.10.2.3	Measure the critical parameters only	39
	1.10.3	Monitoring methods		39
1.11	Concluding observations			40

CHAPTER 2	**ENVIRONMENTAL INVESTIGATION AND MONITORING OF THE FIXED LINKS ACROSS THE DANISH STRAITS**	**41**
2.1	Background	41
2.2	Environmental requirements	42
	2.2.1 Monitoring strategy	43
	2.2.2 Turbidity and sedimentation monitoring programme	45
	2.2.3 The eelgrass programme	46
	2.2.4 The mussel programme	48
2.3	Resulting impact and conclusion	49

CHAPTER 3	**DREDGING IN THE DUTCH & BELGIAN COASTAL WATERS AND THE NORTH SEA**	**51**
3.1	Introduction	51
	3.1.1 Geography	51
	3.1.2 Hydrography	51

	3.1.3	Topography		54
	3.1.4	Seabed morphology		54
	3.1.5	Seabed sediments		56
	3.1.6	Economy		56
3.2	Dredging activities			59
	3.2.1	General		59
	3.2.2	Maintenance dredging		59
	3.2.3	Port extensions		61
	3.2.4	Beach nourishment		61
	3.2.5	Dredging activities for pipelines, outfalls & shore connections		62
	3.2.6	Reclaiming coastal area's		62
	3.2.7	Sand & gravel extraction		63
		3.2.7.1	Shell extraction	65
		3.2.7.2	Environmental consideration about sand and gravel extraction	65
3.3	Dredging equipment			66
	3.3.1	The trailing suction hopper dredger		67
		3.3.1.1	Applied working area	68
		3.3.1.2	Environmental considerations	71
	3.3.2	The cutter suction dredger		75
		3.3.2.1	General considerations	75
		3.3.2.2	Areas of application	76
		3.3.2.3	Environmental considerations	78
	3.3.3	The plain suction dredger		79
		3.3.3.1	General considerations	79
		3.3.3.2	General considerations	80
		3.3.3.3	Areas of application	81
		3.3.3.4	Environmental considerations	82
	3.3.4	The water injection dredger		83
	3.3.5	The bucket ladder dredger		84
		3.3.5.1	General considerations	84
		3.3.5.2	Areas of application	86
		3.3.5.3	Environmental considerations	86
	3.3.6	The grab dredger		87
		3.3.6.1	General considerations	87
		3.3.6.2	Areas of application	88
		3.3.6.3	Environmental considerations	88
	3.3.7	The backhoe dredger		89
		3.3.7.1	General considerations	89
		3.3.7.2	Areas of application	90
3.4	Environmental aspects when dredging			91
	3.4.1	The trailing suction hopper dredger		94
	3.4.2	Cutter suction dredger		96
	3.4.3	Bucket ladder dredger		99
	3.4.4	Backhoe dredger		100
	3.4.5	Grab dredger		100

CHAPTER 4 DREDGING IN UK COASTAL WATERS 105
4.1 History 105
    4.1.1 Later developments 106
4.2 Types of dredging in the UK 108
    4.2.1 Dredging for Ports and Harbours 108
        4.2.1.1 Environmental legislation 108
        4.2.1.2 Capital dredging 108
        4.2.1.3 Maintenance dredging 111
    4.2.2 Dredging for beach nourishment 116
    4.2.3 Dredging for aggregates for construction 117
4.3 Environmental issues 118
    4.3.1 Dredging 118
    4.3.2 Disposal 119
    4.3.3 Beneficial use 120
4.4 Forward look 121

CHAPTER 5 DREDGING IN SPANISH COASTAL WATERS 125
5.1 The Spanish coast 125
5.2 Dredging in Spain 127
    5.2.1 Port dredging 127
    5.2.2 Beach nourishment 128
5.3 Spanish legislation on dredging 129
    5.3.1 Coastal Law 22/1988 130
    5.3.2 Law of Ports and Merchant Navy, 27/1992 130
    5.3.3 Law on the economic regime of and services provided by ports of general interest (Law 48/2003) 130
    5.3.4 Law 6/2001, on the assessment of the environmental impact 131
    5.3.5 European directives 132
    5.3.6 International conventions to protect the environment 132
    5.3.7 Recommendations for the management of dredged material in Spanish Ports 133
5.4 Three examples of environmental actions 133
    5.4.1 Management of dredged material in the Port of Huelva 133
    5.4.2 Specific dredging in the basin of the Port of Sagunto 135
5.5 Barcelona port expansion project. Corrective measures along the coastline 136

CHAPTER 6 DREDGING IN THE UNITED STATES 139
6.1 Introduction 139
    6.1.1 Brief history of Corps of Engineers and US dredging 139
6.2 US environmental laws and acts 140
    6.2.1 Environmental laws 140
    6.2.2 Brief summary of three major US laws or acts effecting dredging 140
        6.2.2.1 MPRSA – Marine Protection, Research, and Sanctuaries Act–Ocean Dumping Act-London Dumping Convention 140

		6.2.2.2	CWA – Clean Water Act	141
		6.2.2.3	NEPA – National Environmental Policy Act	141
		6.2.2.4	Other Environmental Laws, Acts, and Executive Orders	142
	6.2.3	Permitting		142
	6.2.4	Environmental impact statements		143
	6.2.5	Environmental windows		143
6.3	Jones act			144
6.4	US dredging volumes and costs			145
6.5	US dredging companies and equipment			149
	6.5.1	Dredging companies		149
	6.5.2	Types of dredges		151
6.6	Contaminated sediments and capping in US ports and waterways			153
	6.6.1	Equipment		154
	6.6.2	Placement options		155
	6.6.3	Capping		155
6.7	Major US ports and waterways			157
6.8	US options for placement of dredged material			159
	6.8.1	Open water placement		160
	6.8.2	Upland confined placement		161
	6.8.3	Beneficial use		162
6.9	Modeling of dredging and dredged material disposal			163
6.10	Dredging information, research and education			164

CHAPTER 7	**DREDGING IN HONG KONG**			**167**
7.1	Introduction			167
7.2	Hong Kong's natural marine environment			168
7.3	Environmental regulatory framework			170
	7.3.1	Classification and disposal of dredged sediment		170
	7.3.2	Environmental Impact Assessment		171
7.4	Reasons for dredging in Hong Kong			172
	7.4.1	Sand to form new land as reclamations		172
	7.4.2	Mud removal and disposal		174
7.5	Sand dredging			175
	7.5.1	Historical perspective		175
	7.5.2	Management of sand resources		176
	7.5.3	Exploration for sand and examples of some major dredging projects		176
	7.5.4	Fines content of sand: implications for fill specification and dredging		183
	7.5.5	Environmental aspects of sand dredging		183
	7.5.6	Supply of dredged sand from outside Hong Kong		187
7.6	Mud dredging and disposal			188
	7.6.1	Contaminated mud		188
	7.6.2	Uncontaminated mud		191
	7.6.3	Dredging problems caused by unexploded ordnance		194
7.7	Future dredging in the Hong Kong area			194

CHAPTER 8 SINGAPORE DREDGING 201

8.1 Introduction 201
    8.1.1 Meteorological, hydro-graphic and geological conditions 202
8.2 Dredging activities 202
    8.2.1 Land reclamation 202
    8.2.2 Maintenance dredging of navigation channels and ports 202
    8.2.3 Dredging for beach creation/nourishment 202
    8.2.4 Dredging from outfalls of storm drain 204
    8.2.5 Sand mining 204
        8.2.5.1 Material by levelling hills 204
        8.2.5.2 Seabed sand 204
        8.2.5.3 Recycled material from construction sites 204
        8.2.5.4 Marine clay 205
    8.2.6 Disposal of dredged material 205
8.3 Dredging projects 205
    8.3.1 Earlier reclamation works 205
    8.3.2 Reclamation works since 1959 206
    8.3.3 JTC land reclamation projects 206
        8.3.3.1 Tuas View Reclamation 207
        8.3.3.2 Jurong Island Reclamation 207
        8.3.3.3 Tuas View Extension 207
    8.3.4 HDB land reclamation projects 208
        8.3.4.1 South-east coast 209
        8.3.4.2 Pulau Ubin and Pulau Tekong 209
        8.3.4.3 Northeast coast 210
        8.3.4.4 Woodland Checkpoint 211
        8.3.4.5 Kallang Basin 211
        8.3.4.6 Tuas checkpoint 211
        8.3.4.7 West coast 211
        8.3.4.8 Pasir Ris 211
        8.3.4.9 Marina Bay and Tanjong Rhu 211
        8.3.4.10 Pasir Panjang 211
        8.3.4.11 Southern island reclamation 211
    8.3.5 PSA land reclamation projects 211
        8.3.5.1 Changi Reclamation 211
        8.3.5.2 Changi East Reclamation 212
        8.3.5.3 Reclamation of offshore islands 214
        8.3.5.4 Reclamation and infrastructure works at the southern islands 214
        8.3.5.5 Phases of port development 214
    8.3.6 Dredging for navigation channels 215
        8.3.6.1 Dredging of rock shoals at Main Strait 215
        8.3.6.2 Brani terminal: approach channels and basins 216
        8.3.6.3 Singapore harbour: dredging of approach channels, fairways and basins 216

		8.3.6.4	Maintenance dredging at Tanjong Pagar Terminal and approaches	216
	8.3.7	Dredging for pipeline laying		216
		8.3.7.1	Deep tunnel sewerage system	216
		8.3.7.2	Installation of offshore pipelines from Indonesia to Singapore	217
		8.3.7.3	West Natuna transportation system	217
		8.3.7.4	Shell Pandan pipeline laying	218
		8.3.7.5	Powergas submarine pipeline project	218
		8.3.7.6	SMB-trench dredging of pipeline	218
8.4	Impact of dredging on the environment			218
	8.4.1	Impact on coral reefs		218
	8.4.2	Impact on sandy mudflats and seagrass		219
	8.4.3	Impact in Punggol river		220
	8.4.4	Protection measures		220
		8.4.4.1	Environmental constraints	220
		8.4.4.2	Protection measures	220
		8.4.4.3	Measures against illegal disposal	220
8.5	Disputes with neighbouring countries			221
	8.5.1	Dispute with malaysia		221
8.6	Future development			222

CHAPTER 9   DREDGING IN A CHANGING ENVIRONMENT   225

9.1	Abstract	225
9.2	Introduction: environment as an issue	225
9.3	Different cultural perspectives on the values of nature	226
9.4	From perspectives to attitude: views on dealing with nature	228
	9.4.1 Conservation view	228
	9.4.2 Development view	229
	9.4.3 Functional view	230
9.5	Views and public policies	231
9.6	Globalisation, economic growth and the environment	232
	9.6.1 Local Average Income Levels and Environmental Pressure	232
	9.6.2 Regional Average Income Levels and Environmental Pressure	233
	9.6.3 Global Average Income Levels and Environmental Pressure	234
9.7	From "END-OF-PIPE" to systems innovation	234
9.8	A Typical dredging example of the four stages of response	236
9.9	Implications for the dredging sector	238
9.10	Conclusion	239

Introduction

Dredging has been closely linked to coastal development since about halfway the 19th century when harbours and coastal waterways started to be constructed and dredged to facilitate marine trade and fisheries. A notorious example that involved much (unforeseen) dredging was the construction of the Rotterdam Waterway that directly connected the harbours of Rotterdam with the open sea by avoiding the treacherous sandbanks at the mouths of the Maas-Rhine estuary. Also nowadays dredging of Europort and the Rotterdam harbours is still necessary; it contributed to the development of the Dutch dredging industry and of very large seagoing dredging ships. With similar developments in France, Belgium and Britain it led to the large dredging industry that now operates worldwide.

The dredging techniques used in excavating shipping channels and harbours (usually to maintain a required waterdepth) could also be applied in mining aggregate minerals from the seafloor, in putting subaquatic installations in place, and in land reclamation. Thus dredgers have been employed in collecting aggregate minerals (sand, gravel, mud or clay and mollusc shells) for the construction of dikes, dams, jetties and roads, for the foundations of buildings, houses, industrial plants and airfields, and for the construction of artificial islands. Dredged materials have been widely used in the development of urban settlements and industrial plants, not only as aggregate infill, but also in the production of building materials like concrete and bricks. Beach erosion, sometimes the result of coastal construction, was (and is) countered by beach nourishment and the use of dredged aggregates to reconstruct beaches, coastal defenses, and by the production of sand bags and geotextile tubes filled with aggregates to be used as temporary dams.

Dredging for aggregates also extends to the production of sand and the construction of artificial dunes and new wildlife habitats as well as to mining offshore deposits of mineral sands containing tin, gold, titanium, iron or chromium. Nearshore, dredgers have been used for the construction of basements for dams, dikes and jetties, of underwater beach profiles, for submarine shipping channels on the shelf (as off Europort) and for the emplacement of pipelines and subaquatic structures. The varied use of dredgers has led to the development of a variety of dredger types, from small ones for small-sized inshore projects to very large seagoing dredgers for very large projects, with storage of the dredged material on the ship. A maximum depth of dredging the seafloor was reached off the Canadian east coast where offshore oil installations were placed in an excavation at 120–130 meter waterdepth to protect them from damage by icebergs. At present yearly upwards of 100 million metric tons of aggregate sediment is being dredged: the exact figure is not known and fluctuates because very large projects give a temporary large increase.

This book contains chapters on dredging operations in the Netherlands, Belgium, the UK, Spain, the US, Hongkong, and Singapore. Additional chapters discuss more general aspects such as dredging techniques, monitoring of dredging operations, and the prospects for dredging in a changing environment. The latter is especially important, as dredging operations, in particular the larger ones, may have a considerable environmental impact. All dredging to some extent changes the environment, if only through the excavation of a lake or channel bottom or the seafloor, it also affects the environment through dumping of dredged material elsewhere in the water or on land, and by the dredging operation itself, through land reclamation with the dredged material, through associated activities on land or through disturbance because of noise, obstruction of shipping or because of damage to the shore. The expansion of dredging activities, both in area and volume, and the increase of other uses of the aquatic environment have led to conflicts between dredging and other activities and to an increased attention to environmental effects. This caused an increase in rules and regulations concerning the extraction and disposal of dredged material, taking into account possible effects on the subaquatic topography and sediment type, the water quality and the aquatic ecology. In return, this has led to the development of techniques to comply with the regulations in an economic way.

It is the editor's hope that this book may give, besides information on dredging activities in different areas, more insight into the activities and problems (environmental or other) involved in modern dredging.

For more information the reader is referred, besides to the references given in the separate chapters, to "Terra et Aqua", International Journal on Public Works, Ports & Waterways Developments, published by the International Association of Dredging Companies IADC, to the IADC publications, and to the information available under separate headings on the Internet, obtainable through Google or another search programme.

D. Eisma

Chapter 1

Dredging Techniques; Adaptations to Reduce Environmental Impact*

Gerard H. van Raalte
*Boskalis/Hydronamic, Papendrecht, The Netherlands***

1.1 INTRODUCTION

Since the industrial revolution, mankind has interfered with the environment at an ever-increasing rate. Our past lack of understanding and appreciation of our environment have brought us to a point where this interference is manifesting itself in ways that can no longer be ignored.

Studies carried out on a worldwide basis have shown that, if we are to try to guarantee the future of man's existence, we must take a more responsible attitude towards how we live and behave in our environment. Growing public awareness and concern endorse this view.

Dredging is a necessary activity in man's development. It is not a goal in itself but should be considered as a tool to improve and adapt our surroundings to meet the requirements of modern industrial or living standards. However, any human activity will have an effect on the environment, either positive or negative or a combination of both. Dredging and relocation of dredged material are no exception. It is, therefore, of the utmost importance that we should be able to determine whether any planned dredging will have a positive or negative impact on our environment. Evaluation of environmental impact should examine both the short- and long-term effects, as well as the sustainability of the alerted environment.

In its simplest form dredging consists of the excavation of material from a sea, river or lake bed, and the relocation of the excavated material elsewhere. It is commonly used to improve the navigable depths in ports, harbours and shipping channels, or to win minerals

---

* This article is an adapted and abbreviated version of the 4th guide of the series "Environmental Aspects of Dredging", published by IADC and CEDA. It is published with the kind permission of both organisations and of the original author, Mr. J. Smits, of IMDC, Belgium.

IADC (International Association of Dredging Companies. The Hague, Netherlands) is an umbrella organisation for contractors in the private dredging sector. Its objectives are to advance fair trade practices, sound tendering procedures and the use of standard contract conditions; to improve communication about dredging in general; and especially to encourage research into environmentally sound technologies and beneficial uses of dredged materials.

CEDA (Central Dredging Association, Delft, Netherlands) is a member of the World Organisation of Dredging Associations which promotes the interests of all those in the dredging industry and related industries, public or private, individuals or companies.

CEDA organises meetings and conferences for the study of dredging technologies and develops publications for the exchange of information in the field.

** Boskalis/Hydronamic. P.O.box 209, 3350 AE Papendrecht, The Netherlands; phone +31786969099; fax +31786969869; e-mail g.h.vanraalte@hydronamic.nl; www.boskalis.nl; www.hydronamic.nl

from underwater deposits. It may also be used to improve drainage, reclaim land, improve sea defence or clean up the environment.

When using dredging techniques, we must be aware of the environmental effects of the changes we are trying to achieve, as well as the effects of the dredging activity itself. Such changes may include:

- alterations to coastal or river morphology, e.g. enhancement or loss of amenity, addition or reduction of wildlife habitat;
- alterations to water currents and wave climates, which might affect navigation, coastal defence and other coastal matters;
- reduction or improvement of water quality, affecting benthic fauna, fish spawning and the like;
- improvement of employment conditions owing to industrial development; and
- removal of polluted materials and their relocation to safe contained areas.

The dredging activity itself will also affect the environment, often to a lesser extent in the long-term than will be immediately apparent. Environmental effects of dredging may include short-term increases in the level of suspended sediment in the vicinity owing to the effects of the excavation process, the overflow whilst loading hoppers, and the loss of dredged material from hoppers or pipelines during transport and at the disposal disturbance or loss of benthic fauna.

Frequently, the level of suspended sediments generated by dredging activities are no greater than those caused by commercial shipping or bottom fishing operations, or even those generated during severe storms. However, it is often difficult to demonstrate this without undertaking comprehensive studies. Once again, re-suspension of contaminated materials poses special problems and demands rigorous scientific analysis.

An additional positive environmental aspect of dredging, one that is actively encouraged by the controlling authorities, is the beneficial use of dredged material, including some that is contaminated. Typical uses include beach nourishment for sea defence, the creation of wetlands for recreation and wildlife sanctuaries, reclamation of land for commercial and industrial development, and the improvement of agricultural land.

This article concentrates on the equipment and techniques available for dredging and disposal, and evaluates these on basis of environmental performance.

1.2   TYPES OF DREDGING PROJECTS

Before commencing a detailed discussion about the environmental effects of the different types of dredging equipment, it is a prerequisite to fully understand the purpose for which these dredgers are used. Dredging is a general term, covering a wide variety of different activities. However, a review of dredging applications reveals three main groups: capital, maintenance, and remedial works.

1.2.1   Capital dredging works

Capital dredging involves the creation of new or improved facilities such as a harbour basin, a deeper navigation channel, a lake, or an area of reclaimed land for industrial or residential purposes. Such projects are generally characterised by the following parameters:

– relocation of large quantities of material;
– compact soil;
– undisturbed soil layers;
– low contaminant content (if any);
– significant layer thickness; and
– non-repetitive dredging action.

Clearly, a negative environmental effect of such dredging or disposal action is often the destruction of natural habitats (e.g., reclamation of wetlands, disposal of excavated material in biologically sensitive zones, disappearance of inter-tidal flats). However, on the positive side, additional wetlands or inter-tidal flats can be created and important sites can be protected from erosion.

1.2.2 Maintenance dredging works

Maintenance dredging concerns the removal of siltation from channel beds, which generally occurs naturally, in order to maintain the design depth of navigation channels and ports. The main characteristics of maintenance dredging projects are:

– variable quantities of material;
– soft soil;
– contaminant content possible;
– thin layers of material;
– occurring in navigation channels and harbours;
– repetitive activity.

Since maintenance dredging occurs mainly in artificially deepened navigation areas, the dredging activity is, in itself, not necessarily damaging to the natural environment. The main potential for environmental impact is from the disposal of the dredged material and by the increasing quantities of suspended sediments during the dredging process (possibly inducing dispersion of the contaminants). These problems are compounded by the need to repeat maintenance dredging regularly, since siltation is a never-ending story.

1.2.3 Remedial dredging works

It should be appreciated that both capital and maintenance dredging can have a beneficial side effect of removing contaminated material from the dredging location. Remedial dredging is not always recognised as a separate type of dredging, and yet it has distinctive characteristics. It is carried out in an effort by society to correct past actions which have, in some cases, resulted in heavily contaminated sediments. Remedial dredging requires the careful removal of the contaminated material and is often linked to the further treatment, reuse or relocation of such materials. Its characteristics are:

– small dredged quantities;
– high contaminant content;

– soft, uncompacted soil;
– non-repetitive activity (provided the problem is effectively controlled at source).

Given that this type of dredging is aimed at remedying an existing adverse situation, the main environmental effect is bound to be positive, provided the dredging is done with great care and does not significantly damage the environment in any other way. A prerequisite for a successful remediation project is the removal of the contaminant source prior to the start of any remedial dredging.

Clearly, each of these three categories of dredging has different goals, but none has dredging as a goal in itself. Dredging is a means to an end, such as the deepening of a harbour, the removal of contaminated material, the creation of wetlands or the construction of a safe place for industrial or residential development.

1.3   CRITERIA TO JUDGE THE ENVIRONMENTAL EFFECTS OF A DREDGER

The environmental effects of an industrial activity are a complex amalgam of interacting processes in a wide range of different domains. It is difficult, therefore, to compare different dredging equipment and projects with each other on their specific merits. A framework is required to enable us to identify the most significant environmentally sensitive criteria, which may be influenced by the dredging equipment and process.

In a general approach, the following criteria have to be taken into account:

– *Safety of people*: both safety and health of the crew on board the dredger, especially where contaminated sediments are being handled, as well as the safety of all other people.
– *Accuracy of the excavated profile*: important to excavate a pre-set profile as accurately as possible in order to minimise the volume for further treatment or storage.
– *Suspended sediment*: excessive suspended sediments may endanger local fauna and flora.
– *Mixing of different soil layers*: when mixing layers of soil with different geotechnical or chemical characteristics problems of material relocation, treatment or reuse can become complicated.
– *Creation of loose (mobile) spill layers*: destruction of the cohesiveness of the upper soil layers, without complete removal, facilitates later natural erosion processes.
– *Dilution*: dredged fine-grained material mixed with large quantities of water leads to considerable dewatering difficulties at the relocation site, and it increases the volume of material to be treated and thus treatment costs.
– *Noise generation*: where dredging is taking place in a populated area or nearby a nature reserve noise can cause disturbance, depending on distance of source.
– *Normal output rate*: dredgers output is directly related to costs; environmental requirements reducing output will increase costs.

The influence of each of these criteria depends greatly on the nature of each particular project. It is the task of the project sponsor to identify the importance of each characteristic.

In cases where the dredging process is adapted or special equipment is developed in order to reduce the environmental effects of a certain type of dredger, the criteria defined above can also be helpful in evaluating the environmental efficiency.

1.4 PHASES OF A DREDGING PROJECT

The characteristics of the dredging cycle change considerably from one project to another. However, it is possible to identify a number of different phases that are common to almost any dredging project regardless of the type of equipment used for the execution of the project. These phases are:

- disintegration of the in-situ material;
- raising of the dredged material to the surface;
- horizontal transport; and
- placement or further treatment.

In the following section the different phases are discussed briefly, with the emphasis on the environmental effects which can result during each of them.

1.4.1 Dislodging of the material

Dislodging is generally carried out by a cutting device such as cutter head, draghead or the cutting edge of a bucket. This excavation process can be relatively easy in the case of soft sediments, but sometimes, where the removal of hard rock is concerned, it can be very difficult.

The most significant environmental effects occurring during the disintegration process are:

- *Increase of suspended sediments*: The dislodging process breaks the cohesion of the in-situ material and part of the material can be brought into suspension by the rotating or straight cutting movement, depending on the energy level applied to excavation and the way in which the material is raised to the water surface.
- *Mixing of soil layers*: When using equipment designed to cut thick layers (combined horizontal and vertical cutting movement), it is difficult to avoid the mixing of different layers.
- *Dilution*: To facilitate further transport of the disintegrated material, water is mixed with the material during the cutting and suction process. The ratio of soil to water varies from one type of dredger to another.

1.4.2 Raising the material

During the second phase of a dredging cycle, the disintegrated material is raised towards the water surface. This can be done either mechanically or hydraulically. Using the mechanical alternative, the material is raised in a bucket (backhoe, dipper or bucket ladder dredgers) while the hydraulic dredgers (cutter suction, trailing suction hopper, disk bottom, auger and sweep dredgers) use a suction pipe.

The main environmental risks during this raising phase are:

- *Release of suspended sediments*: In the case of mechanical raising in an open bucket, the dredged material is in direct contact with the surrounding water which can result in dilution and an increase in suspended sediment content of the surrounding water layers during the raising process.

- *Loose and mobile spill layers*: With hydraulic transport, potential problems are limited to the point at which the material enters the suction mouth and any residual spill layer of loose material remains on the sea bed. The same effect is observed when the dislodged material falls slowly through the water column, arriving at the suction depth after the suction mouth has moved away.
- *Density of material*: Problems can be encountered with hydraulic transport as this method requires the addition of water to create a mixture density suitable for hydraulic transport. The co-ordination of cutting capacity and pumping capacity is critical to obtain the optimal mixture.
- *Overflow during loading of hopper or barges*: Overflow of excess water inevitably brings sediment into the surrounding waters. This effect is greater when hydraulically loading as compared with mechanical dredging. Prohibition of overflow when dredging soft sediments will effectively prevent loss of sediment into the water column.

1.4.3 Horizontal transport of the material

The third phase of the dredging cycle is the horizontal transport of the excavated and raised material from the dredging area to the site for further treatment or final relocation. This can be achieved by one of two methods, hydraulic pipeline transport or transport by hoppers or barges.

Each method is linked primarily to a certain type of dredger. Barge transport is generally selected for mechanical excavation, while pipeline transport mainly occurs with hydraulic dredgers. Of course, other types of transport, such as truck or conveyor belt transport exist but to date their application in the dredging industry has been limited. In the future these alternative transport techniques should be considered as possibilities, especially for remedial dredging projects.

The environmental effects of the horizontal transport phase can be summarised as follows:

- *Safety*: Using open barge transport, the crew may well come into direct contact with the dredged material, especially of concern in case of dealing with contaminated material.
- *Dilution*: Dilution occurs mainly with the use of hydraulic dredgers where a maximum density is imposed to enable pipeline discharge with a centrifugal pump.
- *Spillage*: The impact of the transport process is slight compared with the other phases of the dredging cycle, only leakage at pipeline joints or spillage from barges or hoppers during transport can be of relevance.
- *Noise and air pollution*: The potential effect on this variable is more significant in the case of barge transport compared to pipeline transport. However should truck transport be used, the effects are greater.

1.4.4 Placement of the material

The final phase of a dredging project is the relocation of the excavated material to its final destination or to an intermediate site for further treatment. There are numerous options at this phase: reclamation of a site, beach nourishment, wetland creation, relocation on land, relocation in a pit, relocation at sea (underwater) or relocation in an isolated site.

A detailed discussion on this subject is outside the scope of this article. Reference is made to the 5th guide of the IADC/CEDA series on "Environmental Aspects of Dredging".

However, some preliminary information on the physical effects of the relocation process is mentioned below, and some special low-impact disposal techniques are discussed in Chapter 8 of this article.

The following effects of relocation can be noted:

– *Occupation of space and surfaces*: The major effect, especially for relocation on land, is the occupation of the ground and the destruction of the natural habitat at that location, but also with underwater relocation, although the effect is less visible.
– *Dispersion of the deposited material*: At underwater relocation sites, depending on water depth, the effects of natural wave and current conditions can result in the dispersion of the fine-grained material into the surroundings during or after the actual relocation action. At land-based relocation sites, wind erosion and dispersion by wind or by rain run-off can occur after the relocated materials have dried. Also finer sediment can be re-dispersed by poorly decanted effluent from placement sites or settlement ponds.
– *Noise and air pollution*: The placement action, especially when the relocation site has to be formed with trucks and other earth-moving equipment, can generate noise and air pollution problems.
– *Groundwater quality*: If the selection of the relocation site is incorrect and the design or construction of protective measures (liners and such) is poor, the groundwater can be affected by leakage.

1.5 THE DREDGING EQUIPMENT

In this chapter the different types of dredging equipment are described briefly, with special emphasis on those characteristics that influence environmental effects as defined by the criteria at Chapter 3. A more detailed description of the different types of dredging equipment can be found in various textbooks on dredging equipment, e.g. in Bray, Bates and Land (1997). Environmental impact of dredging, especially focussed on sediment plumes, can be found in CIRIA (2000).

In Chapter 6.1 new technical developments introduced to avoid or mitigate the negative environmental effects of different types of dredging equipment are reviewed. In Chapter 6.2 new types dredgers, especially developed for low-impact projects, are discussed. Of course other equipment exists which is not included in this article, but in general such equipment is not frequently used in the dredging industry.

1.5.1 Hydraulic dredgers

Hydraulic dredgers include all dredging equipment which makes use of centrifugal pumps for (a part of) the transport process (raising or horizontal transport). Generally speaking, three main groups of hydraulic dredgers can be identified: the Stationary Suction Dredgers (SD), Cutter Suction Dredgers (CSD) and Trailing Suction Hopper Dredgers (TSHD).

1.5.1.1 *Suction Dredger (SD)*
The Suction Dredger is the simplest form of hydraulic dredger. From the floating pontoon the suction pipe is lowered into the bottom and by mere suction action of the dredge pump, often mounted on the suction ladder, bottom material is sucked up (Figure 1.1). Only

8 DREDGING IN COASTAL WATERS

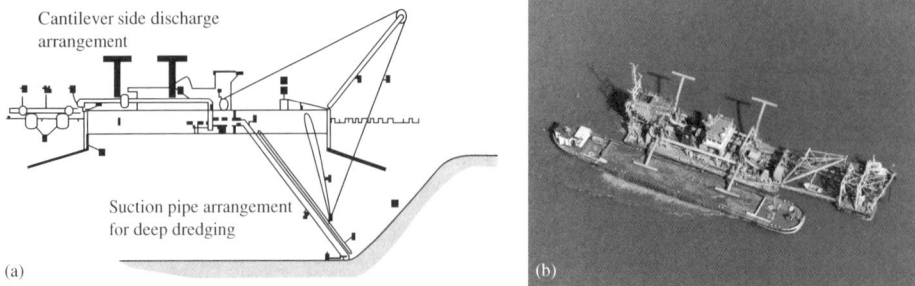

Figure 1.1. (a) Typical section of stationary Suction Dredger. (b) Suction Dredger loading barges.

relatively loosely packed, granular material or silt can be dredged with this equipment. The application of water jets near the suction inlet improves excavation capacities. After raising the material through the suction pipe the material is either hydraulically discharged through a floating pipeline to the shore, but more often loaded into barges.

Regarding the environmental effects of the SD the following can be mentioned:

– *Safety of the crew*: The transport processes occurs within a completely closed circuit. The crew has no direct contact with the material, except when a blockage in suction mouth or pump has to be removed.
– *Accuracy of the excavated profile*: Owing to the relatively uncontrolled production process by suction, normally an irregular pattern of pits is created. Accurate dredging is not possible. Only when the suction entrance is modified with a wide and flat head, a so called "dustpan", a better precision can be obtained, in the order of 10 to 20 cm.
– *Increase of suspended sediments*: Depending on the difference between jet-flow and suction flow the SD has in principle a low tendency to re-suspend sediments. During vertical and horizontal transport, the increase in suspended sediments is non-existent because of the closed pipeline.
– *Mixing of soil layers*: The SD is less suitable for selective dredging, as the production is based on the flow of material from a high bench.
– *Creation of loose spill layers*: The production process of a SD is based on a free, relatively uncontrolled flow of material to the suction mouth. Consequently considerable spill is to be expected.
– *Dilution*: Water is added to the soil for transportation purposes. Depending upon the soil type and the attainable layer thickness, the amount of added water varies significantly.
– *Noise*: As only pump(s) and winches are to be powered for a quite constant process, noise disturbance is low, especially when engines are properly maintained.
– *Output rate*: SD output rates vary widely from 50 to 5,000 m^3/hr depending upon the size of the SD and the soil characteristics. As transport is mainly done by barges, the efficiency is also affected by the size and organisation of the barge transportation system.

The SD is mainly used for sand wining purposes. As the production process is based on a relatively uncontrolled principle of sand flows, the SD is not suitable for environmental sensitive dredging works. Only when working against lower faces and equipped with a "dustpan" the SD might by used for specific environmental tasks.

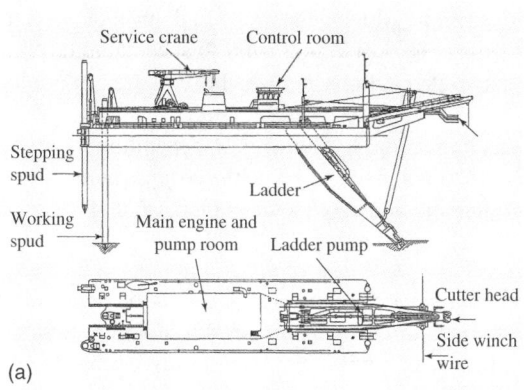 

Figure 1.2. (a) Typical sections of Cutter Suction Dredger. (b) Front view of CSD with cutterhead.

1.5.1.2 *Cutter Suction Dredger (CSD)*

The basic design of the CSD is shown in Figure 1.2. The CSD dislodges the material with a rotating cutter equipped with cutting teeth. The loosened material is sucked into the suction mouth located in the cutter head by means of a centrifugal pump installed on the pontoon or ladder of the dredger. Further transport of the material to the relocation site is achieved by hydraulic transport through a discharge pipeline (partly floating, partly land based). Occasionally the material can be pumped into transport barges for further transport.

Regarding the environmental effects of the CSD, the following can be mentioned:

– *Safety of the crew*: As with the SD, transport processes occur within a completely closed circuit. The crew has no direct contact with the material, except when a blockage in the cutter or pump has to be removed.
– *Accuracy of the excavated profile*: Good accuracy can be obtained, because the movement of the dredging head is controlled from a fixed point (the working spud). Accuracies down to 10 cm are feasible, although at full productivity the accuracy level is approx. 25 cm.
– *Increase of suspended sediments*: The swing speed of the ladder and the rotating speed of the cutter create additional suspended sediments at the dredging site. Careful selection of these parameters is important in order to reduce these effects. During vertical and horizontal transport, the increase in suspended sediments is non-existent because of the closed pipeline.
– *Mixing of soil layers*: For optimal use of the CSD the minimum layer thickness should equal the complete height of the cutter, which mostly is not the case when selective dredging.
– *Creation of loose spill layers*: Most CSDs do not have an optimal combination of cutting capacity and suction capacity for all types of soil. Consequently a spill layer (0.25 to 1 m) remains in general on the sea bed after dredging if no special precautions are taken. An additional pass at the same dredging depth can remove this spill layer.
– *Dilution*: Water is added to the soil for transportation purposes. Depending upon the soil type and the permissible layer thickness, the amount of added water varies significantly.

- *Noise*: Generally the CSD has a powerful engine which generates a high level of noise in the immediate vicinity of the dredger. This noise level diminishes to acceptable levels a few hundred metres from the dredging site. Precautions are possible, but are not implemented on a routine basis.
- *Output rate*: CSD output rates vary widely from 50 to 5,000 m^3/hr depending upon the size of the CSD and the soil characteristics. For a given soil type the cost per cubic metre of the dredging operation with a CSD generally decreases with an increase in the size of the dredger.

The CSD is used mainly for capital dredging in harder soils, which have to be removed in thick layers. The transport distance to the relocation or reclamation site should preferably be limited to allow for pipeline transport. In case of environmentally sensitive projects, the dredging process must be controlled very carefully.

1.5.1.3 Trailing Suction Hopper Dredger (TSHD)

The basic design of the TSHD is given in Figure 1.3. The TSHD is a normal sea-going ship equipped with a suction ladder. At the end of the suction ladder is a draghead, which can be lowered onto the sea bed while the TSHD navigates at a reduced speed. During the forward movement of the TSHD, the draghead dislodges a thin layer of the sea bed. The loosened material, together with some transport water, is sucked into the suction pipe by means of a centrifugal pump, installed in the vessel's hull.

The material is pumped into the ship's hopper until it is completely filled. Then the suction pipe with draghead is retrieved on board the ship. It is also possible to continue loading even though the hopper is filled with a mixture of water and sand. During this loading phase, excess water flows overboard together with some of the finer material, while the coarser (sand) fraction accumulates in the hopper, thereby increasing the quantity of sediment effectively loaded into the hopper during the dredging process.

Horizontal transport is achieved by navigating the ship to the relocation site which is often an underwater site. At this underwater relocation site, the bottom doors in the TSHD's hopper are opened and the sediment falls to the sea bed. As an alternative to such direct

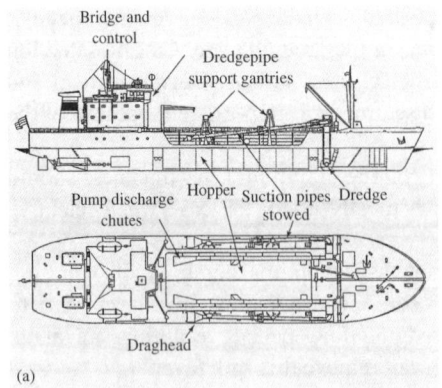

Figure 1.3. (a) Typical sections of TSHD. (b) One of the largest TSHD's in the world.

disposal, many TSHD's are equipped with a system to use the suction pump to empty the hopper by pumping the material through a pipeline to a relocation site on land.

Regarding the environmental effects of the TSHD, the following can be mentioned:

- *Safety of the crew*: Once the dredged material is pumped into the hopper, the crew may come into contact with it. In the exceptional case of a high methane gas content in the dredged material, often arising from port maintenance dredging, the health of the crew can be in jeopardy.
- *Accuracy*: The accuracy of the dredging depth is relatively low, owing to the fact that the position of the suction pipe is flexible and difficult to control. Normal accuracy is around 0.5 to 1 m vertically and 3 to 10 m horizontally. Accuracy can be improved provided sophisticated monitoring and steering equipment is used.
- *Increase of suspended sediments*: The actual cutting process creates little suspended sediments. However, overflow of the excess water and losses of fines cause a significant increase of suspended sediments throughout the water column at the dredging site.
- *Mixing of soil layers*: The cutting process is strictly horizontal. As such, the mixing of soil layers can be controlled accurately. However, the TSHD is not suited to the removal of layered (contaminated) material.
- *Creation of loose spill layers*: Because the dislodging process is basically a scratching and eroding action, only limited amounts of soil are loosened, without leaving a residual spill layer. Larger spill layers can be generated by the settlement of large quantities of overflowing fine-grained material.
- *Dilution*: A certain amount of water is added during the suction process. Depending on soil to be dredged, pumping-ashore needs an additional volume of water with the dredged material for pipeline transport.
- *Noise generation*: The TSHD is equipped with powerful engines generating significant noise levels in the immediate vicinity of the dredger. The noise level is reduced to acceptable levels at a distance of a few hundred metres. A TSHD generally works in more distant areas; as such, noise generation is less critical.
- *Output rate*: Output rates vary widely, ranging from 200 to 10,000 m^3/hr, depending on the size of the TSHD, soil characteristics and transport distance. The cost of a dredging project using a TSHD generally decreases with an increase in the size of the TSHD.

The TSHD is used mainly for maintenance dredging projects or deepening existing channels. During such projects a limited thickness of softer material has to be removed, and relocation sites are at variable distances.

This type of dredger is also used for winning good quality sand far out at sea for reclamation projects such as beach nourishment or the creation of artificial islands. Selection of the optimal duration of the suction process and limiting overflow losses during dredging, are the major factors when trying to control the environmental effects of this type of equipment. For example, stopping the dredging process at an early stage will reduce the overflow of fine material from the hopper. However, this results in higher dredging costs per cubic metre of dredged material. To find an optimal solution, ecological and economic consequences should be evaluated together.

1.5.2 Mechanical dredgers

The second category of dredger is the mechanical type. This includes all plant, which makes use of mechanical excavation equipment for cutting and raising material. Generally speaking, three sub-groups can be identified: bucket ladder dredgers (BLD), backhoe dredgers (BHD) and grab dredgers (GD).

1.5.2.1 *Bucket Ladder Dredger (BLD)*

The basic design and principal components of the BLD are given in Figure 1.4. First employed in Europe, the BLD is the most traditional type of dredger and consists of a large pontoon with a central well in which a ladder, equipped with an endless chain of buckets, is mounted. During dredging, the endless chain rotates along the ladder. The lowest bucket digs into the bed material and the cut material falls into the bucket. It is then carried upwards as the bucket chain rotates. At the upper end of the ladder, the bucket turns upside-down and the soil falls into a chute which guides the material into a barge for further transport.

As far as the environmental effect of the BLD is concerned, the following can be mentioned:

– *Safety of the crew*: The possibility of crew members coming into contact with excavated material is high. This risk exists during the whole process, when the material is being raised in open buckets, loaded into barges and further transported in open barges.
– *Accuracy*: The precision of the BLD is good as the cutting edge of the successive buckets passes at the same depth as long as the ladder remains in the same position. Accuracy to within 10 cm can be obtained. The BLD is, therefore, often used for dredging projects where accuracy is vital.
– *Increase of suspended sediments*: Some additional suspended sediments are released during the raising of the material in open buckets as they move at a relatively high speed through the water. During this raising movement some spillage can occur over the complete height of the water column. This effect can be limited by reducing the speed of the bucket line, which directly affects the output rate.
– *Mixing of soil layers*: The BLD can easily cut relatively thin layers, avoiding the mixing of different soil layers. However, a minimal thickness is necessary for full productivity; otherwise the buckets become partially filled with water.

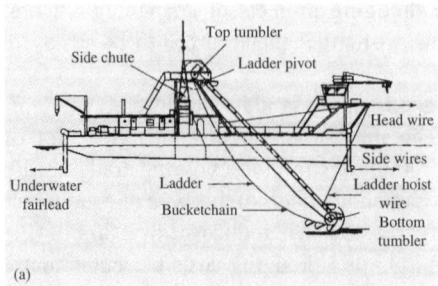

Figure 1.4. (a) Typical section of BLD. (b) BLD at work.

- *Creation of loose spill layers*: Almost all the soil loosened by the bucket is carried away by the rotating bucket chain leaving a clean surface. A minor risk of a spill layer remains if there is excessive spillage while the material is being raised.
- *Dilution*: As the material is raised mechanically, there is no need for transport water. Only when the buckets are not completely filled with soil, quantities of water fall with the soil into the transport barges. The quantity of added water is in any case much less than compared with the hydraulic dredgers.
- *Noise generation*: Owing to the mechanical movement of large steel buckets over a steel framework, the BLD is the worst equipment with respect to noise. For smaller BLDs some trials have been carried out in which new types of material (instead of steel) have been used for the most critical parts. Noise levels decrease to acceptable levels at a distance of a few hundred metres.
- *Output rate*: Output rates of 50 to 1,500 m^3/hr can be achieved.

The BLD is used mainly for accurate dredging such as for tunnel or pipeline trenches. However, taking into account the high density of the excavated material, the BLD is well suited to the excavation of fine-grained material when the addition of transport water can cause problems and if good geotechnical characteristics are required at the relocation site. The raising of material in open buckets and the contact with the water column during this phase of the dredging process are drawbacks to using the BLD for remedial dredging projects.

1.5.2.2 Backhoe Dredgers (BHD)

The basic design and principal components of the BHD are given in Figure 1.5. The BHD is basically a conventional hydraulic excavator, mounted on a pontoon equipped with a spud carriage system. The bed material is excavated by the crane's bucket, which is then raised above water by the movement of the crane arm. The material is placed in a transport barge.

Horizontal transport to the relocation site is generally carried out by transport barges. The material is either deposited through the bottom doors of the transport barge, pumped ashore using a barge unloading dredger, or mechanically unloaded by a grab or hydraulic excavator.

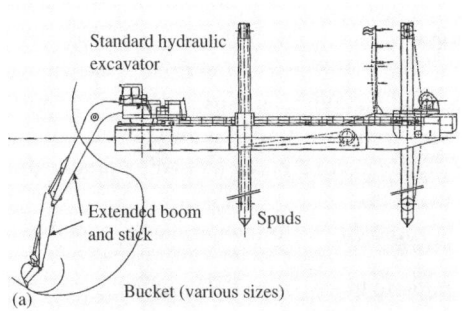

Figure 1.5. (a) Typical section of BHD. (b) BHD at work.

Regarding the environmental effect of the backhoe dredger, the following can be mentioned:

- *Safety of the crew*: The risks encountered with a BHD are similar to those of a BLD in that the material is raised mechanically through the water column and the onward transport by barge is also identical.
- *Accuracy*: The accuracy of the BHD is variable, as the excavating bucket has to be repositioned at every cycle. Without sophisticated monitoring equipment, an accurate dredging depth is impossible. However, such monitoring systems exist and are now implemented on a routine basis for the larger BHDs. Accuracy down to 10 cm is reachable, albeit with reduced productivity.
- *Increase of suspended sediments*: Re-suspension of sediments occurs when the bucket is raised through the water column. During this part of the dredging cycle the operator must give full attention to keeping the bucket in an optimal horizontal position in order to prevent spillage.
- *Mixing of soil layers*: Thin layers can be excavated provided a good monitoring and control system is available.
- *Creation of loose spill layers*: For the same reasons as with the BLD, the BHD does not leave a spill layer.
- *Dilution*: Once again, the BHD is similar to the BLD. Where there is hydraulic unloading of the transport barges, there is a need for transport water.
- *Output rate*: The output rates of the BHD are limited: up to 500 m^3/hr is achievable with the largest BHDs.

The BHD is mainly used for the execution of relatively smaller dredging projects in harder soils as the mechanical cutting forces which can be applied are considerable. Until recently this type of dredger was seldom used for environmentally sensitive projects because it lacks precision and dredged materials are raised in an open bucket.

Recent developments in sophisticated monitoring and control equipment and in new types of closed buckets have improved accuracy considerably, making this type of dredger attractive for more precise dredging projects, and for areas where debris are expected or where physical constraints prevent the use of more traditional equipment.

1.5.2.3 *Grab Dredger (GD)*

The basic design and principal components of the GD are given in Figure 1.6. The GD is basically a conventional cable crane mounted on a pontoon. The bed material is excavated by the bucket of the crane and raised by the hoisting movement of the cable. Once above water the crane arm swings and the material is discharged into a transport barge for horizontal transport. At the relocation site the material is either discharged through the bottom doors of the transport barge or pumped ashore by means of a barge unloading dredger, or mechanically unloaded by a grab or hydraulic excavator.

Regarding the environmental effects of the grab dredger, the following can be mentioned:

- *Safety of the crew*: The risks with a GD are similar to those with a BHD since the material is raised mechanically through the water column and onward transport (by barge) is also identical.

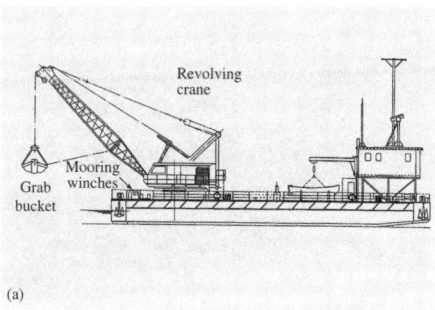

Figure 1.6. (a) Typical section of GD. (b) The largest GD used in dredging.

- *Accuracy*: The accuracy of the GD is limited because the excavating bucket has to be repositioned at every cycle. Without sophisticated monitoring equipment, accurate positioning is impossible. Horizontal accuracy is poor, especially in deep waters and in water currents where a pendulum effect occurs.
- *Increase of suspended sediments*: The problem here is similar to that of the BHD.
- *Mixing of soil layers*: With a traditional GD it is very difficult to achieve a horizontal cut as the excavation depth of each cycle cannot be kept under control. Therefore, mixing different layers cannot be avoided. Recently, new bucket types and monitoring and control systems have been developed with improved characteristics in this respect.
- *Creation of loose spill layers*: Because the GD uses a mechanical cutting process to scrape the material from the bed, no spill layer is left.
- *Dilution*: The situation here is again similar to that with the BHD. Where hydraulic unloading of the transport barges takes place, there is a need for transport water.
- *Output rate*: The output rates of the GD are limited, heavily depending on water depth at the dredging location; at "normal" depths to some 300–800 m^3/hr. However, there are a few huge grab dredgers with considerably higher output rates (1,000–2,000 m^3/hr).

The GD is mainly used for the execution of relatively smaller dredging projects. Recent developments in sophisticated monitoring and control equipment and new types of buckets have improved the accuracy of this dredger considerably. This has made it also attractive for more accurate dredging projects. Furthermore, closed grabs, which prevent direct contact between the excavated material and the water during the raising movement, are available.

1.5.3 Hydrodynamic dredgers

Hydrodynamic dredging is the deliberate (re)suspension of sediment from the sea/riverbed with the aim of removing this material from the dredging area using natural processes for transportation. The water column itself is used as the primary transport medium for the dredged material, instead of pipes, barges or hopper, as with conventional dredging techniques. Reference is made to van Raalte (1999). In this group two main types of dredger are discussed here: water injection dredgers (WID) and underwater ploughs (UWP).

16 DREDGING IN COASTAL WATERS

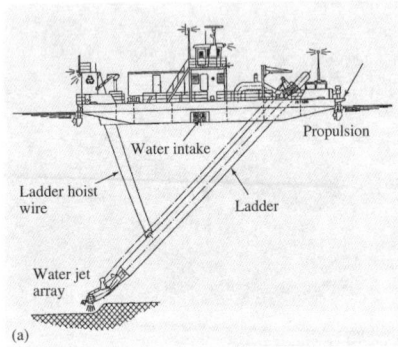

Figure 1.7. (a) Typical section of WID. (b) WID with artist's exploded underwater view.

1.5.3.1 *Water Injection Dredger (WID)*
The basic design and principal components of the WID are given in Figure 1.7. The WID injects high quantities of water into the upper layers of the bed material. Because of the decreased density and the produced higher top level, the material starts flowing naturally until a new equilibrium is reached. In the case of (even small) gradients at the dredging site the transport distance can be significant. The material settles again at an adjacent site of lower elevation.

The following environmental effects can be mentioned:

– *Safety of the crew*: As transport processes occur almost exclusively at the river bed, contact between the crew and the material to be dredged is almost non-existent.
– *Accuracy*: The accuracy of the WID is low as it is very difficult to control effectively the penetration depth of the injected water. Selective or precise dredging is thus impossible with this equipment.
– *Increase of suspended sediments*: During water injection a cloud of suspended material is created at the dredging site, which mostly remains close to the river bed and is therefore less subject to spreading over the full water column.
– *Mixing of soil layers*: This obviously occurs during the injection process, which is difficult to control effectively especially in the softer soil layers.
– *Creation of loose spill layers*: The actual dredging process involves the creation of a loose layer that moves under the natural hydrodynamic forces.
– *Dilution*: Dilution is inherent to the WID process.
– *Output rate*: The output rates of the WID are limited. However, under favourable natural conditions (sloped surfaces and loose natural soil layers) production levels of up to 1,000 m^3/hr can be reached.

The WID can in general not be used for environmentally sensitive projects, because the material is not physically removed from the environment; it is merely shifted to another location in a less controllable, and not always predictable, way. However, for dredging projects in less sensitive areas the WID has significant potential advantages.

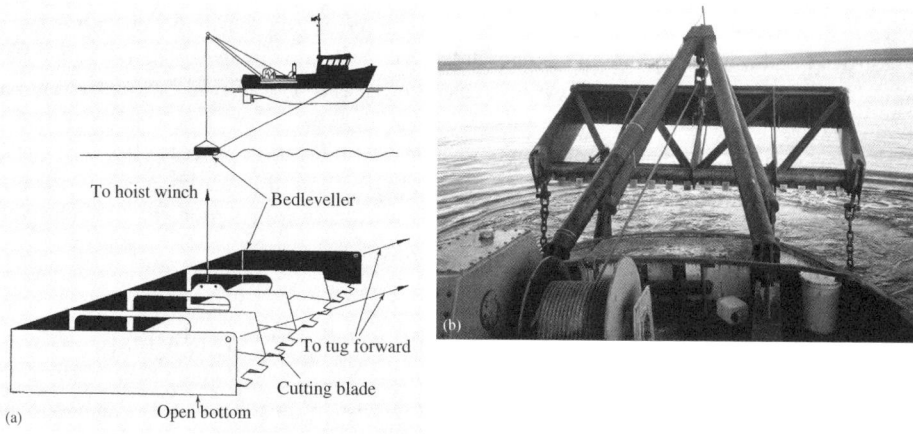

Figure 1.8. (a) Typical scheme of UWP. (b) Tug with UWP.

1.5.3.2 *Underwater plough (UWP)*

The basic design and principal components of the underwater plough are given in Figure 1.8. The UWP can be described as a huge frame that is pulled over the river bed by a tugboat. The frame is equipped with a cutting blade that scrapes over the river bed cutting the bottom layers. The cut material, in front of the cutting blade, is pushed forward until the volume in front of the blade is full. At that moment the transport process is either stopped or the material falls off the blade and goes partially into suspension.

The following environmental effects can be mentioned:

- *Safety of the crew*: As transport processes take place almost exclusively at the river bed, contact between the crew and the material to be dredged is almost non-existent.
- *Accuracy*: The accuracy of the UWP is low. Some control of the penetration depth of the cutting blade is possible by a careful handling of the frame's suspension wires or by using rigid arms.
- *Increase of suspended sediments*: During the cutting process a cloud of suspended material is created at the dredging site, which mostly remains close to the river bed and therefore settles rapidly again. In some cases ploughing is supported by the injection of air from the frame, so as to improve the transport by natural water movement at the dredging site, but also increasing suspended sediment loads.
- *Mixing of soil layers*: As it remains difficult to control the actual cutting depth of the blade, selective dredging is not an option as differentiation between the transport and relocation site is almost impossible.
- *Creation of loose spill layers*: This phenomenon is limited in the dredging area. However, in the relocation area the cut material is left without further handling and, since it has been cut and moved, there will be a significant reduction in its consistency, making it easily erodable.
- *Dilution*: Dilution is not significant with the UWP, unless air injection is applied.
- *Output rate*: The output rates of the UWP depend mainly on the size of the cutting blades. The largest blades are now approx. 20 metres wide and up to 2 metres high. These UWP dredgers can reach high output rates (up to 2,000 m^3/hr).

The UWP is mainly used for maintenance dredging in tidal basins where significant quantities of natural sedimentation accumulate. The material is pushed back to the natural water current in the main channel, or to areas where it can be removed further by regular dredging.

Because it is impossible to control the transport and relocation process effectively, and because the material is not removed from the natural system, this dredging method is not suitable for environmentally sensitive projects. However, for other projects the technique can be an attractive alternative because of the low cost.

1.6 RECENT DEVELOPMENTS IN LOW-IMPACT DREDGING EQUIPMENT

The worldwide increase in environmental consciousness in recent decades is also evident in the dredging industry. As awareness grew that some parts of sediments which must be removed are more or less charged with contaminants, mainly originating from the pollution of the surrounding waters, so did concerns about potential risks. And because of the contaminant content in the dredged material, it became more and more difficult to find suitable reuse or disposal options for the material.

Listening to the general concerns of the public, the responsible authorities or sponsors and the dredging industry concluded, in some cases jointly, that normal practices and procedures had to be changed, especially where contaminated sediments are involved. This understanding forced the dredging world to rethink its role. It was deemed necessary to:

1. reconsider maintenance dredging in industrialised ports, taking into account the potential contaminant content of the sediments;
2. develop measures to overcome negative environmental effects other than by ceasing the dredging operation;
3. adapt existing dredging equipment in order to reduce their environmental features;
4. initiate monitoring and control procedures to measure in real time the environmental effects of the dredging activities;
5. initiate studies to gain a better understanding of the contamination characteristics of the sediments and the potential risks linked to these contaminants;
6. initiate a series of laboratory and prototype tests to evaluate new dredging, treatment and disposal techniques for contaminated sediments;
7. reconsider the selection of the most appropriate dredging equipment for certain dredging projects, taking into account the presence of contaminated sediments or other environmental constraints;
8. develop new types of dredgers, better adapted to the requirements of the revised tasks.

In this chapter on low-impact dredging techniques and equipment items 3 and 4, on development of existing tools, and item 8, on development of new tools, are discussed. In all other items the dredging industry intensively co-operated in developments initiated and conducted by port operators, national authorities, institutes and other stakeholders. Reference of this interaction might be found in other articles in this book, and in the IADC/CEDA series on "Environmental Aspects of Dredging".

1.6.1 New developments in existing dredging equipment

Besides the development of almost completely new equipment especially designed for remedial dredging (described further in Chapter 6.2 of this guide), a significant effort has been put into the further development and optimisation of existing dredging equipment.

During the last decade, considerable change has taken place with respect to the constraints imposed on dredging activities, particularly in response to environmental considerations and limitations.

The main developments all aim at improvement of the environmental criteria as discussed in Chapter 3, and have led to a large number of new systems and equipment being installed on board the different types of dredgers, several of which are described and discussed below.

1.6.1.1 *Degassing system, for hydraulic dredgers*
A factor which limits the density of the transport mixture pumped by centrifugal pumps is the gas content of the bed material. In order to overcome this limitation it is necessary to reduce the gas content of the material before the sediment/water mixture enters the suction pump.

In recent years various systems have been developed aiming to extract (a significant part of) the gas content just in front of the suction pump. This results in significant increases in density in the transport pipeline or in the TSHD's hopper. This in turn results in reduced relocation volumes.

1.6.1.2 *Overflow reduction, for TSHD*
One of the major environmental constraints of a TSHD is the suspended sediments generated by the overflow of excessive transport water with a high content of fines. One obvious way to overcome this problem is to stop the dredging action when the hopper is full. However, this results in a rather uneconomical dredging cycle. Therefore new technologies have been developed, such as:

- low-density trailers that have a relatively large hopper well, with better features for the settlement of dredged material in the hopper;
- controlled overflow to improve the retention capacity of the dredger;
- controlled overflow, directing the overflow water back to the river or sea bed, and thus reducing the spread of suspended sediments into the surroundings considerably;
- reuse or recirculation of overflow water in the draghead of the dredger. This drastically reduces the excess water that is discharged freely overboard during a dredging cycle;
- use of submerged pumps on the suction pipe of TSHD's, which permits a higher density of material to be dredged and which therefore reduces the need to overflow by achieving a higher hopper load.

1.6.1.3 *Turtle deflecting device, for TSHD*
A specific development is the turtle deflection device (Figure 1.9), developed by the US Army Corps of Engineers. Mounted on a TSHD draghead, it is used to avoid the entrapment of sea turtles (an endangered species) through the suction pipe of a TSHD during maintenance dredging in Florida (USA).

1.6.1.4 *Dredging Information Systems, for all types of dredgers*
During recent years there has been a spectacular improvement in the monitoring of dredging activities, both on board vessels using a continuous data logging system as well as in the

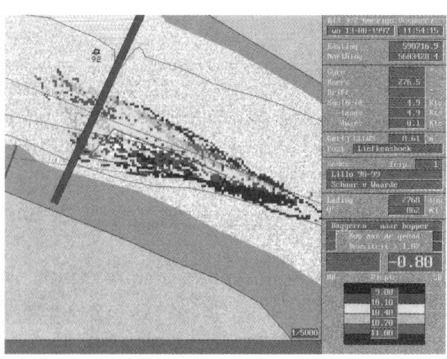

Figure 1.9. Turtle deflecting device.

Figure 1.10. DIS system used on Western Scheldt, Belgium.

office, using Dredging Information Systems (Figure 1.10). These systems continuously register a large number of variables, which are interpreted automatically and supplied to either a control and steering system that is virtually automatic, or, at least, to a real time feedback system for use by the work supervisor and dredge master. The development of profile monitors, depth indicators, dynamic tracking systems and multi-page monitors are a few visible results of this trend.

The main result of the use of Dredging Information Systems is a tighter control of the dredging depth and quantities, which allows for a decrease in the tolerances for maintenance and capital dredging works. This diminishes the volumes to be dredged for a given dredging project and makes it feasible to remove (contaminated) material in a layer-by-layer approach.

1.6.1.5 *Monitoring and control systems for mechanical dredgers*
The most significant development to improve the accuracy of the mechanical dredging equipment, especially the BHD and the GD, is the continuous computer-based monitoring of the position of the bucket underwater. Traditional mechanical monitors have gradually been replaced by electronic monitors, which give the dredge master accurate information (up to 5 cm) on the actual position of the cutting edge of the bucket. The more sophisticated systems memorise the dredging history, thus helping the dredge master to avoid dredging the same place unnecessarily (Figure 1.11).

1.6.1.6 *Positioning systems for mechanical dredgers*
A major aid is the use of significantly more accurate positioning systems, providing within centimetres the actual cutting depth and position. This is achieved by using laser equipment or kinematics DGPS systems, sometimes installed on the ladder or excavating arm of the dredger, enabling instantaneous measuring the movements of the cutting edge and correcting the position either automatically or manually.

Regarding the environmental effects, this system, in combination with the improved monitoring capability (6.1.5), allows for a much reduced tolerance. This has improved the possibility of excavating contaminated material at almost in-situ density, a basic characteristic of the mechanical dredgers, without unacceptable over-dredging.

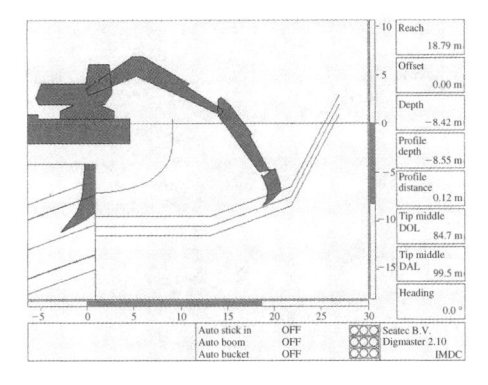

Figure 1.11. Screen of backhoe monitoring system.

Figure 1.12. Visor dredging grab.

1.6.1.7 *Special buckets for BHD and GD*

As far as grab dredgers are concerned, a new type of bucket has been developed to overcome the inaccuracy of the penetration depth at each cycle. The new buckets have a specially designed closing system which keeps the cutting edges of the bucket at a constant depth.

Another development for BHD and GD is the closed bucket or grab (Figure 1.12). This reduces direct contact between the excavated sediments and the surrounding water to a minimum when raising bucket or grab.

1.6.1.8 *Encapsulated bucket line for BLD*

To prevent re-suspension of sediments by washing off and spillage of soils form the open buckets of a BLD, while they are raised over the ladder, a completely encapsulated ladder has been developed (Figure 1.13). With this new BLD the bucket line is raised in still water (no currents around the buckets) and, should some spillage occur, this spill remains in the encapsulated area and is guided in a natural way along the ladder down to the river or sea bed.

1.6.2 Development of new dredging equipment

Especially in the context of remedial dredging, rather spectacular developments have taken place during the last decade. In this period a complete new range of dredging equipment has been developed. The design of this new equipment is based on existing technology, but takes into account the overall goal of improving the performance of the equipment with regard to criteria (see Chapter 3) that are critical to an ecologically sensitive task, namely, removal of contaminated (but also non-contaminated) sediments, without charging the environment unnecessarily.

In the following pages, some of the newly developed dredgers are described in more detail, to be considered as "current state of the art" (year 1998). It can however be expected that within a few years, once developments have been tested and applied on full-scale environmental dredging projects, new developments will emerge, providing even further improved equipment and output rates.

Figure 1.13. BLD with encapsulated bucket line.

Figure 1.14. Cylindrical shaped disc bottom cutter.

1.6.2.1 Disc bottom cutter dredger

The disc bottom cutter is a classic stationary dredger equipped with a cylindrical shaped cutter with a flat closed bottom and a vertical rotation axis. The suction mouth for removal of the cut material is situated inside the cutter to avoid spillage. A shield over the full cutter height, at places where no soil is encountered, and an automatically adjustable shield above the soil to be cut, prevent both the cut material from entering the surroundings, as well as the intake of excessive water volumes (see Figure 1.14). A flow rate pump-control system and a degassing system are installed on board the dredger to improve the density of the mixture during the pipeline discharge process.

The disc bottom cutter dredger features a number of environmental improvements when compared with more traditional dredgers:

- *Safety of the crew*: The basic arrangement is a closed-circuit hydraulic pipeline system. The health and safety risks for the crew are therefore minimal.
- *Accuracy*: The accuracy of the disc bottom cutter dredger is basically the same as that of a stationary dredger, which can position its cutting edge within centimetres of the target depth. Combined with the specially designed flat bottom cutting device, an accuracy of less than 5 cm can be achieved.
- *Suspended sediments*: The completely closed shield around the cutter of the disc bottom cutting device is designed to avoid the spread of the cut material. The creation of suspended sediments is therefore minimal and limited to the near surroundings of the cutting device.
- *Mixing of soil layers*: The automatic steering device makes it possible to excavate different layers selectively.
- *Creation of loose spill layers*: The closed shield around the cutter prevents the material cut by the dredger from escaping through the suction mouth. Consequently, the residual spill layer is minimal or non-existent.
- *Dilution*: The disc bottom cutter dredger remains a hydraulic dredger, which requires a minimal quantity of transport water. To reduce the quantity of undesired process water during unproductive stages of the dredging process, the disc bottom cutter dredger is

Figure 1.15. Close-up of sweep head with movable visor.

equipped with a system that automatically slows down the flow in the discharge pipeline to the critical transport velocity.
– *Noise generation*: Noise generation is similar to that of other CSDs.
– *Output rate*: The output rate is somewhat restricted compared with a traditional CSD, because most effort is not put into optimising the output rate, but into reducing negative environmental effects, such as creation of additional suspended sediment content, low accuracy and mixture density. Output rates of up to 500 m^3/hr can be achieved with the existing equipment.

The disc bottom cutter dredger is a powerful tool for the execution of environmentally sensitive, remedial dredging projects where accuracy within acceptable budgets is required. The technique has been in use for approx. 10 years. As such this equipment has passed the experimental stage, although continuous upgrading takes place based on project experience. Where the material to be dredged contains debris (such as plastic bags, chains, and stones), the cutting device frequently can become blocked because of the relatively small suction openings of the disc bottom cutter.

1.6.2.2 *Sweep dredger*
The sweep dredger (see Figure 1.15) is based on a classic stationary dredger equipped with a sweep head which is similar to the draghead of a TSHD. A movable visor makes it possible to operate with this type of suction head in two opposite directions, while the cutting height can be optimised continuously. During the dredging process numerous parameters are monitored and controlled by the highly sophisticated steering and control system.

The sweep head, equipped with a visor, shaves using the lower cutting edge the designated soil layer as defined during the preparatory survey, while the upper visor precisely follows the bottom profile in order to prevent the inflow of excess water. The swing speed is controlled automatically according to the pre-set hourly production rate, while the pump speed is fixed to permit hydraulic transport of the cut soil with a minimum volume of additional water. The dredger is also equipped with a sophisticated degassing system.

Compared with more traditional dredgers, the sweep dredger introduces a series of improvements:

- *Safety of the crew*: The sweep dredger is basically a hydraulic dredger with a completely closed circuit. The health and safety risks for the crew are consequently minimal.
- *Accuracy*: Given that the sweep dredger is based on a traditional hydraulic stationary dredger, accurate steering of the cutting edge is possible. This combined with the sweep head design, means that an accuracy of less than 5 cm can be achieved if the dredger is used carefully.
- *Suspended sediment*: The sweep head is basically similar to the draghead of a TSHD, in that it contains no rotating devices, which could generate re-suspension of sediments around the sweep head.
- *Mixing of soil layers*: The automatic steering device allows the excavation of different pre-defined layers selectively.
- *Creation of loose spill layers*: The cutting edge continuously shaves the material to be excavated, guiding the material into the sweep head and suction mouth. In combination with an optimal control of the suction pump this prevents the creation of a spill layer.
- *Dilution*: The sweep dredger requires the addition of water for hydraulic transport through the pipeline, although the automatic steering system is programmed to keep the quantity of additional water to a minimum.
- *Noise generation*: Noise generation is similar to that of other CSDs.
- *Output rate*: The output rate is close to the output rate of a traditional CSD, up to 1,200 m^3/hr can be reached for a full layer thickness, while the environmental criteria can still be met.

The sweep dredger is a powerful tool for the execution of environmentally sensitive, remedial dredging projects where accuracy within an acceptable budget is sought. The technology is new but based on well-known concepts. As such, it is to be expected that the prototype stage will soon be passed.

The efficiency of the sweep dredger can decrease rapidly when the consistency of the material increases, since no active cutting device is provided to reduce lumps in the suction mouth.

The basic design of the sweep dredger calls for the separation of the excavation and vertical raising phase from the horizontal transport phase by the introduction of a (small) buffer basin, aiming at a further reduction in volume of unnecessary transport water during the unproductive stages.

1.6.2.3 *Environmental auger dredger*

The environmental auger dredger (see Figure 1.16) is specially designed for the removal of thin layers of contaminated sediments. The dredger is a normal stationary dredger equipped with an auger that cuts the material in layers with a thickness ranging from a few centimetres to one metre. The thickness of the layers being cut can be maintained continuously within this range. The screws in the auger transport the material to the centre, where a dredging pump sucks away the material through a suction mouth. The suction force and a screen or skirt around the auger prevent dispersion of the material into the surrounding water.

Figure 1.16. Environmental auger dredger.

The width of the cut depends on the width of the auger. A sophisticated monitoring and control system is installed to optimise environmental efficiency. A degassing system is fitted to the dredger to maximise the density of the mixture during the pipeline discharge process.

Compared with more traditional dredgers in respect of environmental criteria, the environmental auger dredger offers a number of improvements:

– *Safety of the crew*: The cutting and pumping system of the environmental auger dredger is completely enclosed. Thus the risks of direct contact between the crew and the transported material are minimal.
– *Accuracy*: The environmental auger dredger is based on a stationary dredger with a spud system. This allows, in principle, an accurate positioning of the cutting head, which, combined with the automatic steering and control system, allows the auger dredger to work to tolerances of less than 5 cm.
– *Suspended sediments*: The auger is completely closed off from the environment by a skirt which covers the cutting opening. Combined with the rather slow rotating movement of the auger, this results in little suspension of sediments around the cutting device.
– *Mixing of soil layers*: The sophisticated automatic steering system allows for accurate layer-by-layer excavation. Only when there is a rapid three-dimensional variation of the inter-layer boundary, does the length of the auger cause some difficulties in following small-scale variations in both directions at the same time. However, such small-scale variations are not often encountered in nature and it is very difficult to measure them during the pre-dredging survey.
– *Creation of loose spill layers*: The auger cuts and conveys the material towards the suction mouth of the dredger, which is located at the centre of the auger. This feature, combined with good control of the pumping process, eliminates the spill layer.

Figure 1.17. Cable arm grab.

- *Dilution*: The environmental auger dredger requires the addition of transport water. It is very similar to the sweep dredger and the disc bottom cutter dredger, which means that with a good monitoring and steering system, the amount of water needed can be reduced to a minimum.
- *Noise generation*: Noise generation is similar to that of other CSDs.
- *Output rate*: The output rate is determined by the size of the auger and is generally lower compared with a traditional CSD, as most of the effort is aimed at reducing the environmental effects of the equipment. Output rates up to 800 m^3/hour can be achieved.

The auger dredger is a powerful tool for remedial dredging projects where accuracy and environmental effects are of prime concern. The equipment has been in use for approx. 5 years and can be considered passed the experimental stage, although ongoing up-grading is necessary to deal with continuously changing demands. The large dimensions of the cutting device, which limit the number of passes to be made, are profitable when a large area has to be dredged at high precision. When a more variable three-dimensionally target surface has to be followed, a wide auger is less suitable to produce high accuracies.

1.6.2.4 *Environmental grab dredger*
The environmental grab is a specially designed grab with the following features:

- during the opening and closing of the bucket, the cutting edge remains on the same horizontal plane;
- the opening and closing of the grab is undertaken hydraulically;
- when the grab is closed all openings are sealed to minimise spill;
- the crane, operating the grab, is equipped with a high accuracy positioning system.

The environmental grab can be installed either on a traditional grab dredger (cable crane), where it is suspended from cables (see Figure 1.17), or on a backhoe dredger (hydraulic excavator) (see Figure 1.18). The latter, the horizontal profiling grab, enables a better positioning and guidance of the cutting edge during the excavation as the pendulum effect can be avoided.

Figure 1.18. Horizontal profiling grab.

Compared with the more traditional dredgers in relation to environmental criteria, the environmental grab dredger offers a number of improvements:

– *Safety of the crew*: Excavation and barge loading are done mechanically. Although efforts have been made to optimise the process, considerable possibilities remain for direct contact between the crew and the dredged material at different stages of the dredging cycle. Detailed discharge procedures are necessary to avoid careless loading of the barges and the consequent risk of significant splashing.
– *Accuracy*: The accuracy of the cable crane version is reasonable, provided that the most modern technologies are applied. Accuracy levels of approx. 10 cm can be achieved vertically. Horizontal accuracy is less as a result of the free suspension of the grab which causes a pendulum effect, especially in deep waters and in currents. When the environmental grab is mounted on a hydraulic crane, horizontal accuracy improves drastically, achieving levels similar to those of a backhoe dredger.
– *Suspended sediments*: The generation of additional suspended sediments is low compared with normal grab dredgers. However, similar problems can occur as described above under "safety of the crew".
– *Mixing of soil layers*: Clearly, an advanced monitoring and steering system is necessary in order to control the lowering and cutting movement of the grab and the penetration depth, and thus achieve the required accuracy at each intersection of the different layers. The limited accuracy level in the horizontal plane for a crane mounted grab is a drawback in this respect.
– *Creation of loose spill layers*: Taking into account the mechanical characteristics of the cutting process, creation of spill layers can be avoided to a large extent.
– *Dilution*: The dredging process is mechanical and further transport is generally undertaken by barges. Compared with hydraulic dredgers, this offers considerable advantages as far as dilution is concerned. Only when the bucket is partially loaded will some additional water be discharged into the transport barge.
– *Noise generation*: Noise generation is similar to that of a traditional GD or BHD.
– *Output rate*: The output is basically determined by the size of the bucket, and by the time it takes for each bite. As environmental dredging requires higher precision operation, especially in grab positioning, the output rate of an environmental grab is lower than for a conventional grab (GD or BHD), in general limited to a few hundred cubic metres per hour.

Figure 1.19. Pneumatic dredger.

The environmental grab dredger is a mechanical device, which has significant advantages in respect of density of the excavated material. For other environmental criteria, the grab dredger is less advantageous compared with the other systems described. It should, therefore, generally be used in combination with other protective measures. It is suitable for the removal of small quantities at sites that are difficult to reach and for projects where the costs for dewatering or treatment of the excavated material are high in proportion to the volume of material.

1.6.2.5 *Pneumatic dredgers*
For more than a decade pneumatic dredgers such as the Pneuma (see Figure 1.19) have been developed at different locations throughout the world. The system is based on the hydrostatic pumping principles. A differential pressure is induced in a cylinder by a vacuum and the external hydraulic head. This creates an influx of the soft sediments into the cylinder. When the cylinder is filled, the inlet valve is closed and compressed air is pumped into the system and forces the sediments through an outlet valve into the delivery pipeline. When the cylinder is almost empty, the pressure in the cylinder is released, the vacuum is applied again and the entrance valve is opened ready for a new cycle.

The system is similar to a positive displacement pump and has the major advantage that no moving parts, except for the valves, come into contact with the material to be dredged.

Regarding the environmental criteria, the Pneuma dredger has a number of advantages compared with more traditional dredgers:

– *Safety of the crew*: The system is completely closed with very low risk of direct contact between the crew and the dredged material.
– *Accuracy*: Normally the Pneuma system is operated from a cable crane which does not allow for accurate positioning. However, the system can be mounted on the ladder of a stationary dredger, making accurate dredging possible. Position control of the

suction mouth is possible to within 5 cm, while the exact location of excavation is less controllable.
- *Suspended sediments*: As there are no rotating or moving elements, the creation of turbidity is almost non-existent.
- *Mixing of soil layers*: Until now no advanced steering and control system for the cutting depth and positioning has been applied. However, it will be feasible to develop a similar system which would permit a better positioning of the cutting edge and suction mouth.
- *Creation of loose spill layers*: As the movement of the material to be dredged is initiated by pressure differences, there is no spill layer left after dredging.
- *Dilution*: Given that the dredger is based on the positive displacement pump system, only small quantities of transport water are required. However, precautions should be taken to prevent additional water from entering the suction mouth along with the sediments to be dredged.
- *Noise generation*: Noise generation is limited as there are no moving parts. Noise levels are the same as for a cable crane or a stationary dredger, depending on the type of pontoon selected for the project.
- *Output rate*: Output rates of between 40 and 100 m^3/hr have been reported for conventional Pneuma dredgers. However, it is not clear which output rates can be achieved under strictly controlled environmental conditions.

Pneumatic dredgers are attractive for remedial dredging projects because of their high density pumping method, the closed circuit principle, and the fact that no moving elements are in direct contact with the material to be removed (reduced suspended sediment generation). However, the system is vulnerable to debris in the dredging area (a common problem during removal of contaminated harbour sediments) and the automatic steering and control system is far less developed compared with other systems.

It should be mentioned that the system has been developed for the removal of siltation behind hydraulic dams in fairly deep waters. The efficiency with which the material enters the cylinders depends on the pressure difference which is determined by the water depth at the dredging site. With the removal of contaminated sediments, the water depths are generally rather limited, which puts a constraint on this type of dredging equipment for remedial projects.

1.7 TRANSPORT AND DISPOSAL EQUIPMENT

In this chapter the different methods of transport and placement of dredged sediments are discussed. Special attention is paid to those techniques and equipment that can be used to mitigate the environmental effects of transport and placement activities. For each of the transport modes, new developments and possibilities to improve the environmental characteristics are discussed.

1.7.1 Pipeline transport

Pipeline transport (see Figure 1.20) is basically an environmentally friendly transport mode. Compared with other transport modes, energy consumption is low and noise and

Figure 1.20. Pipeline transport.

air pollution are almost non-existent. This generally applies to the transport of dredged material whether it is contaminated or not.

Pipeline transport is also a safe and clean method because it takes place in a closed circuit system. Between entry point at the dredging site and outlet point at the relocation site, there is no possibility for contact between the transported material and the outside world unless a pipeline failure or leakage occurs (which is unlikely provided proper maintenance procedures are followed). Impact of minor leakage can in most cases be ignored and, in any event, it is far less significant than the potential risks from most other methods of transporting dredged material.

To improve the pipeline transport system in this respect, the following developments are being applied:

- Automatic control systems that have been developed include a pump monitoring and steering system for a smoother discharge process with less fluctuation. Given that the upper values are limited by physical laws, this development results in an increase of the average density in the pipeline.
- The use of high-density pumps reduces considerably the need for additional transport water. However, the available types of high-density pumps have rather limited output capacities. This is a major drawback for their application in the dredging industry.
- In cases where the pipeline transport involves the use of a barge unloading dredger, returning the transport water from the relocation site to the dredger and recycling the transport water for the unloading process can be applied. In the case of contaminated sediments, such a procedure is environmentally and economically beneficial.

Pipeline transport is the most environmentally friendly transport method in the dredging industry. The only major disadvantage is the requirement to mix the excavated material with transport water. This increases the volume for storage and/or further treatment, which in the case of contaminated fine-grained sediments can be a serious issue.

1.7.2 Barge transport

The second means of transport frequently used in the dredging industry is hopper or barge transport. In this case the dredged material is loaded onto a ship either hydraulically (TSHD) or mechanically (BLD, BHD, GD). Horizontal transport between the dredging site and the relocation or treatment site is done by navigation of the barge. Barge transport is environmentally friendly with limited environmental impact as regards noise generation, the emissions of exhaust gasses and the creation of road blockages.

The main advantage of this means of transport is the lack of need of transport water. Excavated material can be transported at almost its original density and consistency, provided this density can be maintained during the disintegration process (except for the TSHD which dredges the material with the use of hydraulic pumps).

A minor disadvantage of this means of transport is the fact that most of the barges are open and that the risk of spillage is slightly greater compared with pipeline transport. Furthermore, there is a continuous risk of contact between the crew and the dredged material, which can be a problem in the case of heavily contaminated sediments or sediments with a highly volatile content.

New developments and options for improving the characteristics of this means of transport focus primarily on the limitations previously mentioned:

– *Spillage of material from the barges during transport*: This can be avoided either by placing a cover over the hopper during transport (something as simple as a canvas cover) or by continuously leaving sufficient freeboard (50 cm minimum) in the hopper above the loaded sediments. A third possibility is to allow the sediment to settle for a period of time after finishing the loading operation, after which the water on top can be pumped off, before further transport takes place. Until now neither of these options for removal of excess water has been applied frequently as both have financial implications.
– *Unloading procedure*: Either the barge is unloaded via its bottom doors discharging at an underwater relocation site (a rather uncontrolled procedure not requiring the addition of transport water) or the barge is unloaded using a hydraulic barge unloading dredger resulting in significant dilution during the suction process phase. Alternatively, barge unloading can be made by a mechanical unloading dredger, avoiding dilution of the material, but necessitating a further transportation by a high-density transport mode.

A limitation of barge transport is the prerequisite that the dredging site and the relocation site be linked by a navigable channel with sufficient water depth. If this is not the case, part of the transport must be executed in another way which is most often pipeline transport. In that event, the dredged material has to be rehandled by a hydraulic dredger (barge unloading equipment) with a consequent increase in dilution and volume.

1.7.3 Road transport

Although pipeline and barge transport are used for almost every dredging project, alternative transport modes should be considered, especially if one wants to reduce the overall environmental effects of the dredging cycle, in particular when there is a need for further transport after unloading the barges (i.e., when the destination is not located in the immediate vicinity of a waterway).

The first alternative transport mode is road transport by means of trucks. The main advantages of this method are:

– Trucks can be loaded mechanically at any density.
– Choice of destination is flexible, which is a major advantage when different qualities of material have to be transported to different relocation or treatment sites.

The main disadvantages of this method are:

- Tipper trucks, which are difficult to make spill-free against leaking fluid, are normally used.
- Considering the normal output of a dredging project, the number of trucks necessary for transport of dredged material is high.
- The environmental effects of road transport are greater compared with pipeline transport (e.g., noise generation, exhaust gases, road usage, spillage on public roads, and so on).

Opportunities for the use of road transport in the dredging industry are limited, only acceptable for the transport of dredged material in certain specific cases. However, the trend to dewater fine-grained dredged material (either mechanically or naturally) before final (beneficial) use, places new demands on the transport process and this method can certainly not be ignored.

1.7.4 Conveyor belt transport

A fourth possibility for transporting dredged material on a large scale is the conveyor belt. The system is not commonly used in the dredging industry as the installation costs of such a system are high. Furthermore, the basic characteristics of a wet dredging process are not compatible with the mechanical characteristics of the conveyor belt. However, as the boundary conditions of dredging projects continue to change drastically, all possible transport processes and their applicability to certain types of dredging projects, especially for remedial works, need to be examined.

The application of conveyor belt transport to dredging offers a number of advantages to the industry:

- Mechanical loading of dredged material with no need to add transport water.
- A continuous transport system capable of conveying large volumes of material.
- Transport costs are reasonable for larger volumes.
- Environmental effects (e.g., noise, exhaust gasses, and such) are relatively low, but special precautions need to be taken to reduce noise and avoid dust.

Disadvantages of the conveyor belt transport are:

- Alignment is fixed; changes to this alignment during the transport process are difficult and costly.
- Special precautions have to be taken to avoid losses of dredged material with normal water content.
- Spread the transported material at the destination site can be complicated.

From the above it is evident that the application of this transport mode in the dredging industry is limited to few specific cases. However, there are potential advantages, making it worthwhile considering for use in future projects, especially when there are environmental concerns.

1.7.5 Combined transport cycles

Another aspect of dredged material transport is the increasing complexity of the process, which regularly leads to bi- or even tri-modal transport procedures where two or three different transport modes are used to reach the final destination site (see Figure 1.21).

The following combinations are possible:

- Barge transport combined with a barge unloading dredger that pumps the unloaded material towards the final destination site;
- Barge transport with mechanical unloading and conveyor belt or truck transport to the final destination site;
- Trailing suction dredger with hopper transport and pump ashore facilities (pipeline discharge);
- Pipeline discharge to an intermediate treatment installation and further truck, barge or conveyor belt transport.

Selection of the optimal transport cycle has to be planned with great care as a bi-modal transport process not only combines the advantages of both transport modes but also the disadvantages. As such, it has to be taken into account that each transfer between two transport modes creates the risk of material losses or other environmental risks. Therefore, during the environmental effects analysis of a project, it is necessary to consider the different phases of dredging, transport and treatment or relocation in an integrated way.

1.8 PLACEMENT TECHNIQUES

Placement of the dredged material at the disposal site is another major phase in the dredging process which can potentially have significant environmental effects. Selecting the most appropriate placement site and considering the infrastructure at that destination site are of major importance. In addition, the equipment and techniques used for the placement of dredged material also have an influence on the overall environmental effects of dredging. A discussion on this and on the potential effects of the various placement options follows.

Specific options of what to do with contaminated material are discussed in PIANC (1996).

1.8.1 On-land placement

One option for the placement of dredged material is on land, within a confined area surrounded by dikes. This is generally applied when the use of dredged material is required on land, or in case contaminated material has to be stored on land.

Discharge pumps on board the dredgers are used to pump the dredged material through a pipeline which ends in the confined destination area (see Figure 1.22). The most significant environmental effects of an on-land placement site are the burial of (environmentally sensitive) surfaces, the change of the topography, and the leakage of (contaminated) transport water into the subsoil layers (reference is made to Guide 5 of the IADC/CEDA series "Environmental Aspects of Dredging").

Potential environmental risks during the actual placement action are the overflow of the material and rupture of the surrounding dikes. Both can result in serious damage and

34   DREDGING IN COASTAL WATERS

Figure 1.21. Combined transport: barges, mechanically unloaded onto conveyor belts.

Figure 1.22. Dredged material placed on land in a confined area surrounded by dikes.

spreading of the deposited material into unwanted areas. Proper design of the dikes and regular monitoring of the water level at the relocation area can help avoid these problems.

Another critical item is the evacuation of the excess transport water. Given the large volumes of water which have to be evacuated, this can create serious environmental damage if not properly managed. Both quantity and quality of this effluent water may cause concerns along their discharge route.

Proper design of the disposal area, to ensure maximum opportunity for the material to settle within the area, reduces this effect considerably. Installation of decanter-type basins can be an additional safety measure, where strict limitations are imposed on the suspended sediment content of the evacuated water.

Detailed studies of the natural conditions at and around the planned relocation areas are needed in order to evaluate potential risks and to take necessary precautions, ultimately relocation of the destination site to another less vulnerable place.

1.8.2   Underwater placement

Another option is to place dredged material in open waters (see Figure 1.23). This is mainly done with clean or slightly contaminated material, if suitable disposal sites are available. Underwater placement is generally done after hopper or barge transport where the barge or hopper (TSHD) sails directly from the dredging area to the placement site. At that site the vessel's bottom doors are opened and the material falls from the hopper onto the sea or river bed.

Again, the choice of site has a major effect on the overall environmental consequences of the project. A site with large tidal or other currents generates a greater risk for erosion, re-suspension and further dispersion of the materials into the surroundings. This is not discussed this article. Reference is made to Guide 5 of the IADC/CEDA series "Environmental Aspects of Dredging".

Equipment and techniques used for placement can, however, be adapted to reduce the environmental effects of disposal. Instead of directly opening the bottom doors, the hopper can be emptied by means of a pump linked with a vertical pipeline reaching down to the river

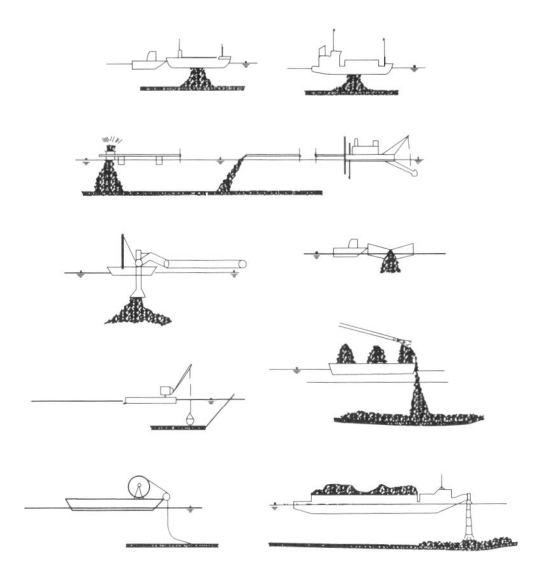

Figure 1.23. Conceptual illustrations for various methods of underwater placement.

Figure 1.24. Underwater diffuser.

or sea bed. In this way the material is guided to its final placement depth without intermediate contact with the overlaying water layers. Losses of fines and dispersion during the fall of the dredged material from the surface to the riverbed or seabed are reduced considerably.

To further optimise this procedure, an underwater diffuser can be fitted to the lower end of the discharge outlet (see Figure 1.24). The diffuser aims at minimising re-suspension of the material by changing the flow from a powerful vertical downward movement towards a gentle horizontal flow just above the sea bed. A correct design of the actual diffuser is crucial. This system can also be applied on larger TSHDs, where one of the suction pipes can be used it in the reverse direction as a kind of fall-pipe, possibly fitted with a diffuser in stead of draghead.

Other means for reducing the environmental effects of the placement process can be realised by restricting the period during which placement is allowed. For instance, avoid underwater placement when maximum tidal currents occur or during seasons when there is intensive biological activity in the area. Construction of underwater bunds or utilisation of underwater pits are other ways of reducing the dispersion of material from the relocation site after actual placement has occurred.

1.8.3 Capping techniques

In the case of underwater disposal of contaminated material, it may be necessary to isolate the material from the environment by means of a capping layer. Such layer protects the contaminated material from erosion by natural water currents and prevents, or at least reduces, the uptake of contaminating elements by the aquatic life, as well as the migration of those elements to the overlying water layers.

Capping is defined as the controlled, accurate placement of contaminated dredged material at an open water site, followed by a cover or cap of clean isolating material (Palermo, 1994). Guides 3 and 5 of this IADC/CEDA series "Environmental Aspects of Dredging" present more details about the requirements for such capping systems. Some of the placement methods as shown in Figure 1.23 can also be applied for cap construction.

1.9 MITIGATING MEASURES

Properly designing a dredging-related project (including conducting an Environmental Impact Analysis) and selecting the most appropriate dredging and placement equipment are undoubtedly the major actions for reducing the overall environmental effects of a project. Still there are other steps that can be taken to reduce these effects.

Such mitigating measures can primarily be taken within the actual dredging process, by provisions on board of the dredgers, during transport and on disposal equipment, as discussed in Chapters 6 through 8. If these are expected, or proven, not to be sufficiently effective, other measures can still be considered.

1.9.1 Measures at the dredging site

Mitigating measures can be taken at the dredging site itself. Apart from careful planning and control of the dredging actions, implementation of physical barriers to prevent the spread of suspended sediments is an important option. This can in some cases be achieved by the installation of silt screens at or nearby the dredging site. The following options exist:

– *complete enclosure of the dredging equipment with a silt screen*: this can only be done with stationary dredgers, using pipeline discharge methods; in other cases the surface to be enclosed is too large or curtains have to be fitted with "sluices to let barges in and out", which is not very practical nor effective.

Figure 1.25. Artist's impression of silt screen at grab dredging.

- *complete enclosure of the dredging zone*: this can be done around the excavation area of grab or backhoe operations, but allowing barges free access to lie alongside and to be changed without hindrance (see Figure 1.25).
- *protecting a sensitive area nearby the dredging site*: in this case the dredger works freely unhindered by the curtain. The curtain is installed in such a way that suspended sediments cannot pass by the curtain towards the sensitive area.
- a combination of both options.

Based on the physical conditions of the site and the environmental restrictions at the location, the type of silt screen, method of deployment and anchoring system can be selected. The installation of such a physical barrier is often a difficult operation, demanding great skill and experience on the part of the dredging contractor, in order to avoid problems of leakage through the curtain.

The use of a silt screen, however, clearly limits the output rate of the dredger, lengthens the execution period, and increases the costs of the project.

1.9.2 Measures at the relocation site

Also mitigating measures at the placement site can be implemented. Again, both planning and physical measures are possible. The following physical measures can be considered:
- installation of a silt screen around the underwater relocation site or around the outlet of a confined placement area;
- utilisation of underwater diffusers to reduce the suspended sediment content at the placement site;
- application of settlement ponds at the outlet of a confined placement area, in order to reduce the suspended sediment content in the excess transport water that is returned to the natural water courses.

In the operational field, the following measures can be considered:
- seasonal restrictions on placement at certain locations;

– tidal restrictions for underwater placement;
– use of absorbent or impermeable liners at the bottom of confined placement areas.

The choice of the most appropriate measures depends largely on the actual conditions at the relocation site. A careful analysis of these conditions is a prerequisite to defining the correct infrastructure and procedures for an optimal project both in terms of economics and environment.

1.10 MONITORING AND CONTROL OF THE DREDGING PROCESS

Proper selection of the most appropriate dredging and/or disposal techniques for a given project is, of course, of major concern when attempting to perform environmentally acceptable work. However, with each possible solution dredging and disposal activities must be controlled and monitored, so as to evaluate the results and effects of the equipment and execution methods selected.

This chapter briefly describes the objectives and planning aspects of a monitoring program relevant to a dredging and disposal process. More details on monitoring can be found in Guide 3 of the IADC/CEDA series "Environmental Aspects of Dredging".

1.10.1 Objectives of the monitoring activities

A programme to monitor or control the environmental effects of a dredging process should be based on compliance, verification, feedback, and know-how as described below.

1.10.1.1 *Ensuring compliance with restrictions*
A major objective in planning a control and/or monitoring programme is to ensure that the dredging process is executed in accordance with the various restrictions, which are legally or contractually imposed. Restrictions can vary markedly from one project to another. They can be either physical (dredging depth, location or transport mode), seasonally related or quality oriented.

1.10.1.2 *Verifying project conditions*
A second objective of a monitoring programme is verification of the hypotheses made during the project preparation. Making these verifications is critical, especially during the first phases of the project's execution, in order to check the validity of the assumptions used as a basis for the project planning and the environmental impact assessment.

1.10.1.3 *Providing feedback*
A monitoring programme is by no means only intended as post factum control of a (dredging) project, to be used as a basis for applying penalties when parameters and criteria are not met during the actual execution. To optimise environmental effectiveness it is of equally great importance that a monitoring programme provides as much direct feedback as possible to the project management team. This will enable them to adjust, wherever possible, the working procedures to achieve even better environmental effects.

Such feedback is not only useful for the project management team, but also for the crew on board the dredger, as they are the key people in the overall success of an environmental protection plan.

1.10.1.4 *Increasing know-how*

One objective of any monitoring programme is to increase the knowledge about the environmental conditions and effects of a given dredging process. This knowledge serves as a basis for a better assessment of the environmental effects during future dredging projects. The procedures below should be considered.

1.10.2 Planning a monitoring programme

Owing to the very large number of parameters involved, a monitoring programme can be time-consuming and costly. In order to limit the amount of energy and money spent on such control, it is important that a monitoring programme be carefully planned, based on the clearly defined objectives of the project.

1.10.2.1 *Define the critical activities*

At an early stage in the project, the potential environmental effects of each piece of equipment on site, possibly sub-divided according to each phase of the project, should be defined. During the planning stage and initial phase of the monitoring and control programme, the most critical equipment and project phases, which generate the highest environmental effects, should be identified. This enables the monitoring team to reduce the extent of the monitoring and control programme during the bulk of the work and to focus, as much as possible, on the critical items of the project. Clearly, the assumptions made during the preparatory phase should be controlled by point check measurements during the initial stage of the actual monitoring campaign.

1.10.2.2 *Take an overall view instead of point measurements*

Generally speaking, the exact location of the greatest environmental effects (e.g., suspended sediment generation) during a particular project is difficult to predict. Therefore, monitoring measurements should be conducted, in as far as possible, over the complete area where a significant effect can be expected.

Most of the parameters to be monitored are of a stochastic nature. Long-term measurements or simultaneous measurements over a larger surface or perimeter are necessary to assess the stochastic variation of a monitored parameter. Therefore, it is advisable to carefully consider the type of equipment which can generate sufficient data for such an assessment.

1.10.2.3 *Measure the critical parameters only*

Environmental effects can be found in a wide range of fields. One standard procedure which controls every potential parameter and location rarely works and is certainly too expensive to implement. Therefore, during the preparatory studies and planning phases, an environmental impact assessment should be carried out in order to define the vulnerable elements at and nearby the dredging and relocation sites, as well as to define the parameters that can be influenced by the dredging process. From these parameters, the most critical should be selected in order to define an ecological and cost-efficient monitoring programme.

1.10.3 Monitoring methods

How monitoring is to be executed, is beyond the scope of this article. For further information reference is made to Guides 3 and 4 of the IADC/CEDA series "Environmental Aspects of Dredging".

1.11 CONCLUDING OBSERVATIONS

The environmental effect of a dredging project, both adverse and beneficial, must be clearly identified at the planning stage. It may largely depend on the type of project: capital, maintenance, or remedial works should be distinguished.

The real environmental effect is a combination of the effects of the selected equipment, the chosen execution method, the design of the project and training, expertise and attitude of the operator. Mitigation of the environmental effects needs to address all three parameters.

The environmental impact of a certain type of dredging equipment has to be viewed in a broad perspective. A significant disturbance over a short period of time can be preferable to less significant disturbances over longer periods.

A project-specific environmental impact assessment is a prerequisite for proper evaluation of the real effect of dredging operations and for proper judgment of the mitigating measures that are worthwhile economically.

Apart from making technical adaptations to the dredging equipment in order to reduce the negative environmental effects, it is recommended that consideration also be given to operational measures such as tidal or seasonal restrictions.

A well-planned and well-executed monitoring and control programme is necessary to ensure that the predicted effects are really met and that mitigating measures actually fulfil the requirements.

In general it can be stated that equipment and techniques and procedures for an environmental sensible dredging project execution are in existence. However, not all of the equipment, and especially the modern special equipment, is available all over the world. Particularly in less developed regions, costs to develop or mobilise such specific tools can be prohibitive. In those areas mitigation of effects can often be achieved with expert dredging advice, modification of existing equipment and operator training.

It must be emphasised that there will always be an environmental impact of some nature and this should be viewed in relation to the benefits gained by executing the project.

REFERENCES

Bray, R.N., A.D. Bates and J.M. Land, *Dredging; a Handbook for Engineers*, Arnold Publishing. London, UK. 1997.

CIRIA C547, *Scoping the assessment of sediment plumes from dredging*, CIRIA, London, 2000.

Palermo, M.R., *"Placement techniques for capping contaminated sediments"*, Dredging 94 Conference Proceedings. Volume 2. US. 1994.

PIANC, Working Group 17. *Handling and treatment of contaminated material from ports and inland waterways CDM.* Supplement to Bulletin no. 89. Permanent International Association of Navigation Congresses. Brussels, Belgium. 1996.

Raalte, G.H. van, and R.N. Bray, *Hydrodynamic dredging*, CEDA Dredging Days Proceedings, Amsterdam, Netherlands. 1999.

Series 'Environmental Aspects of Dredging', *Guide 3: 'Investigation, Interpretation and Impact', 1997, Guide 4: 'Machines, Methods and Mitigation', 1998, Guide 5: 'Reuse, Recycle or Relocate', 1999*, IADC/CEDA, The Hague, Netherlands.

Chapter 2

Environmental Investigation and Monitoring of the Fixed Links across the Danish Straits

Anders Jensen
Chief engineer, DHI Water & Environment, Hørsholm, Denmark

2.1 BACKGROUND

The impact on the marine environment caused by establishing and constructing the fixed links across the Danish straits, the Great Belt and Øresund, was from the very beginning focused on the blocking effect on the water flow between the North Sea and the brackish Baltic. Maintaining the inflow of salt and oxygen rich water from the North Sea to the brackish Baltic is considered by all countries surrounding the Baltic to be very important for the ecological system in the Baltic including the economical important stock of cod.

The environmental concerns about the local environment were much less for the Great Belt Link than for the Øresund Link. The Great Belt Link was planned and constructed during the eighties where the concern about environmental impact caused by large scale construction projects were less than during the late nineties, where the Øresund Link were planned and build. The local marine environment is furthermore less vulnerable in the deep

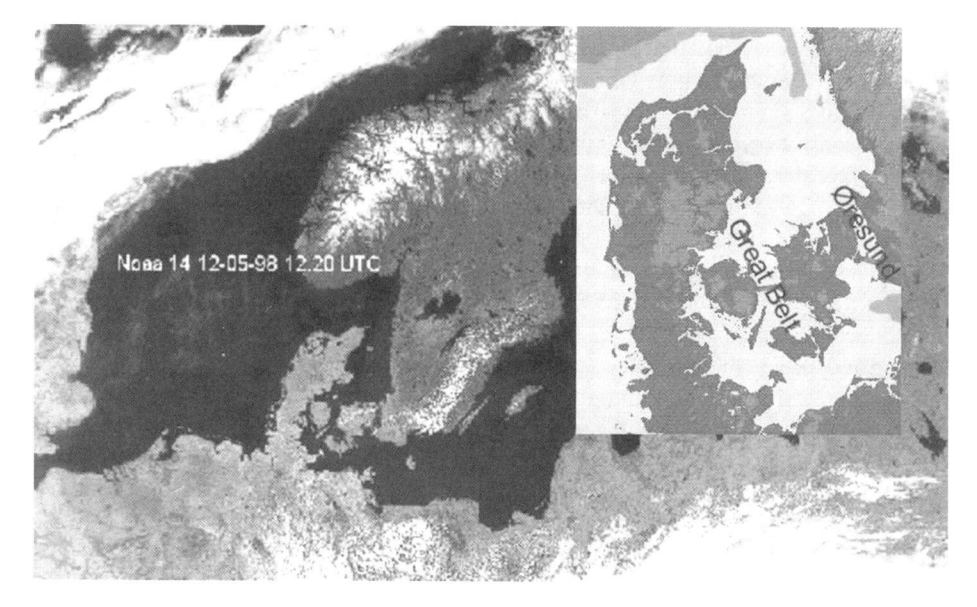

Figure 2.1. The Danish straits Øresund and Great Belt.

and box shaped Great Belt, than in Øresund, where the shallow and clear water gives ideal conditions for the large eelgrass meadows. Øresund is also an important migration route for the Baltic stock of herring.

This chapter will focus on the environmental investigations and monitoring carried out in connection with the construction of the fixed link across Øresund. This project has included all environmental related aspects of a large marine construction project (dredging, landfill and various construction activities) and the environmental concerns and restrictions were focused on the regional environment as well as the local environment. This project was in many ways innovative in the way the very strict environmental restrictions and requirements were tackled.

2.2 ENVIRONMENTAL REQUIREMENTS

The impact on the marine environment caused by establishing a fixed link between Denamrk and Sweden across Øresund was from the very beginning focused on two main concerns:

The blocking effect of the link on the water flow between the brackish Baltic and the North Sea.

The effect on the local environment caused by the dredging and landfill activities.

The water quality of the Baltic is dependent on the water exchange through the Danish Straits. The transport of oxygen and salt from the North Sea is essential, for example, for maintaining the cod stock in the Baltic, an economically important asset. The restriction of the water flow due to the increased resistance to the water flow caused by the presence of the artificial island, peninsula, and bridge piers, has been compensated for (in the Great Belt as well as in the Øresund) by increasing the water depth of carefully selected areas around the fixed Link.

The water quality in the Øresund has greatly improved during the last 20 to 30 years as a consequence of the large investments made in sewage treatment plants by Denmark and Sweden. The water in the Øresund has a low natural turbidity, which results in good visibility under water (more than 10 m during winter) and hence allows bottom flora species such as eelgrass to grow in water depths of more than 8 m. The low lying island of Saltholm provides favourable conditions for water foul such as swans and eider ducks, in addition the southern part of the island is protected by the Ramsar Convention. The Øresund is also a major recreational area for over 2 million people living in the region.

The above-mentioned concerns have lead to the formulation of a set of strict environmental criteria by the Danish and Swedish authorities. Based upon the information gained from the collection of base line data since 1992, it was possible to establish a set of fast reacting and measurable variables, which could be used as feedback monitoring variables.

Environmental monitoring traditionally uses methods that need a long period of observation before one can judge with statistical certainty whether a development is a lasting change or an occasionally occurring natural variation. In connection with the construction of the Øresund Link, an environmental monitoring programme has been established which allows a much quicker evaluation of impacts in order to make adjustments to the construction activities as observed effects follow or vary from predictions.

This so-called "feedback monitoring" includes selected variables that over short periods of time show quantifiable change as a result of impacts from the construction work. The use of computer models makes it possible at an early stage to assess whether a feedback action should be taken or not, given the results of the monitoring and the future work plans.

2.2.1 Monitoring strategy

Environmental Impact Assessments have shown that the largest environmental impact would be due to the dispersal of spilled sediment. Consequently, a great effort has been put into organising the environmental monitoring to ensure that the criteria on the spill have been met.

Two major tools were introduced to ensure that the spill was kept below the limits fulfilling the objectives and criteria for all variables:

The Contractor was made responsible for ensuring that the amount of spill remained below specified limits which varied according to time and place, taking into consideration the environmentally sensitive periods and areas. These limits had been calculated in advance through numerical modelling.

A feedback monitoring programme was implemented by the Owner, covering the dispersal & accumulation of sediment spill and biological key variables representing the important influenced ecosystems.

In theory the contractor's fulfilment of the criteria of maximum spillage alone should reduce the impacts to acceptable levels. However, due to the unpredictability of the hydrographical & meteorological regime and natural variations in the ecosystem, it was necessary to monitor the actual state of the environment during the construction works to ensure the fulfilment of the general environmental criteria.

Feedback Monitoring therefore included selected variables that over short periods of time showed quantifiable change due to impact from the construction work. These variables were measured either continuously or very repetitively and constituted the main instrument for a highly responsive regulation of the construction works.

To ensure the fulfilment of the environmental objectives and criteria, procedures were established to specify the actions to be taken on the construction work where criteria were in danger of being exceeded.

The feedback monitoring served as an integral part of the environmental management system and imposed changes and/or restrictions on the marine works. The feedback procedures included:

- Assessment and approval of equipment, work plans etc. prior to the initiation of operations.
- Application of threshold criteria and feedback loops with an agreed code of action.
- A clear definition of the responsibilities of the involved parties.
- Planning and environmental approvals of the dredging activities were seen as the primary way of ensuring compliance with the environmental requirements.

Thus the key factors for making decisions on actions changing the construction works, were the measurements of the environmental variables combined with forecasts of the future conditions.

In the monitoring programme the use of computer models was integrated in the feedback system. The models were applied during the planning of the dredging and reclamation operations. They were used to forecast the sediment dispersion, sedimentation, impact on

```
                New work plan from the contractor  ←─────┐
                   including a spill scenario            │
                              ↓                          │
                Environmental impact assessment of the   ↑
                    work plan based on numerical
                     modelling of the spill scenario
                              ↓
                Are the operational criteria fulfilled  - NO →
                              ↓
                      The dredging work starts
                              ↓
     ┌─────────→  Monitoring of selected variables
     │                        ↓
     │                                                    ↑
 ← NO -    Are the operational criteria exceeded?
                            YES
                              ↓
                      Intensified monitoring
                       Numerical modelling            ─────
                              ↓
     ← NO -   Are the authorities criteria exceeded? - YES →
```

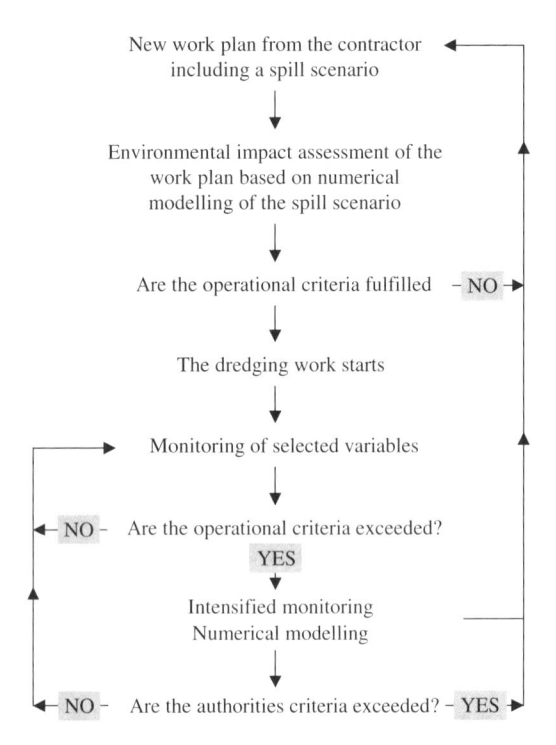

Figure 2.2. Principals of feedback monitoring.

water quality and impact on the eelgrass beds. The use of the models made it possible at an early stage to assess whether a feedback action should be taken or not, given the results of the monitoring and future work plans.

The principles of the overall feedback procedure are illustrated in Fig. 2.2

In the Øresund, eelgrass meadows and mussel banks dominate the plant and animal communities. These were therefore chosen as the most suitable organisms for feedback monitoring in the Øresund. In feedback monitoring it is essential that the variable chosen for measurement meets certain basic demands:

- It must have an unambiguous, easily measurable relationship to the feedback organism which represents the ecosystem concerned.
- The measurement result has to be available in a short time (no more than a few days).
- Background material must be available for the determination of statistically reliable limit values and criteria for judging any limits being exceeded.
- The impact of various conditions on the variable should be calculable in advance, that is, some kind of model of the relationship between cause and effect has to be available.

Given these demands, the following variables were selected for the monitoring of the feedback organism eelgrass:

1. shoot density
2. leaf and root biomass
3. carbohydrates dissolved in the rhizomes.

Figure 2.3. Light attenuation instrument from GMI Denmark used for measuring low level turbidity (0–30 FTU) in Øresund.

Turbidity (sediment concentration) and sedimentation (rates and total) were measured weekly to give early warning data and data to validate the computer model. The selected variables in the mussel monitoring were:

1. distribution
2. biomass.

2.2.2 Turbidity and sedimentation monitoring programme

When dredging and reclamation work started, turbidity and sedimentation surveys were carried out several days a week. Turbidity is a light measurement that allows for the calculation of the concentration of sediment particles in the water.

The surveys served two main purposes. The monitoring provided data for the tuning (calibration and verification) of the sediment model, so that it reflected the real conditions in the Øresund as closely as possible. Also data on turbidity and sedimentation rates were needed to directly control the work carried out in accordance with the operational criteria. The amount of spilled sediment that is transported around in the Øresund waters may be measured in different ways: Either as turbidity measured as light attenuation, (see Fig. 2.3) or as optical back scatter. In some cases sediment plumes may be mapped using acoustic methods (Acoustic Doppler Sediment Profiler). In addition, the sediment fluxes are estimated on the basis of results obtained from sediment traps, (see Fig. 2.4) deployed at several stations in the Øresund.

Figure 2.4. Sediment traps deployed in vertical strings for measuring vertical fluxes of particles.

The transport of sediment along the seabed and sedimentation, are further evaluated on the basis of a combination of submarine video recordings, dual frequency echosounders and sidescan sonar recordings (see Fig 2.5).

The turbidity and sediment monitoring took place in the areas most affected by the spilled sediment. The grey-white plumes of mostly small lime particles could be observed on the sea surface from the research vessel. Occasionally, satellite pictures and aerial photographs were included to support the planning of the surveys. The submarine video recording surveys were planned according to the results of the hindcast modelling of sedimentation, they did not follow a regular pattern, but covered areas where the deposition of spill was expected.

2.2.3 The eelgrass programme

Every second week during the eelgrass growing season (March to November) samples of eelgrass were taken in the area around the Øresund Link. The density of the eelgrass meadows, the weight of the live parts of the plants and the storage of energy in the underground parts were recorded. Samples were taken by divers at ten locations. The locations were chosen according to the ongoing dredging and reclamation work and on the basis of the hindcast modelling. As a part of the set-up and calibration of the numerical model used for calculation of the maximum load of spill, a large field experiment was carried out including shadowing experiments of eelgrass as seen on Fig. 2.6.

Some other items were also monitored to supplement the eelgrass programme. At two sites in the Øresund, light sensors and data loggers were permanently installed on poles

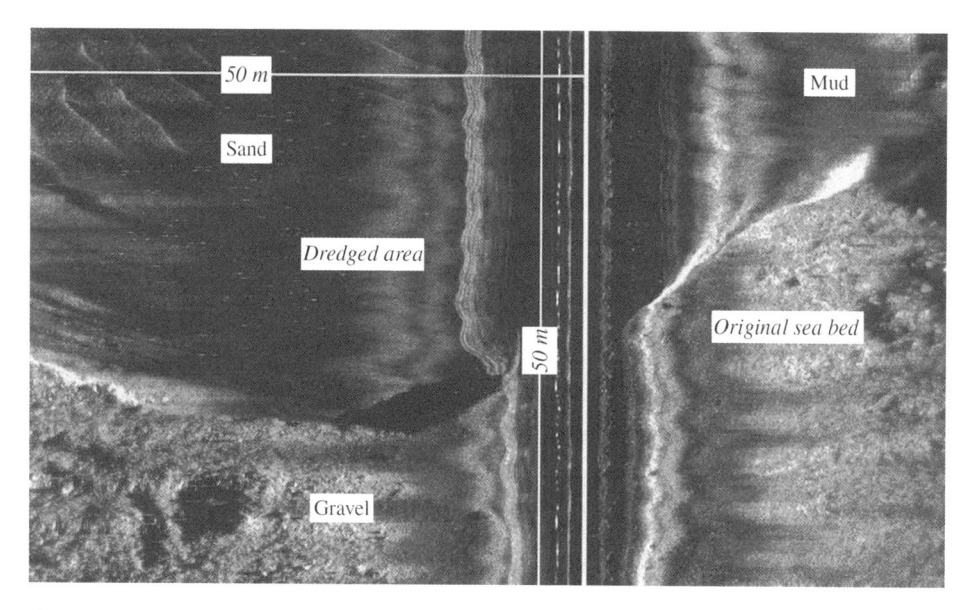

Figure 2.5. Side Scan Sonar image from a dredged area located south of the western end of Peberholm.

Figure 2.6. Eelgrass test area, which is covered to illustrate the shadowing effect caused by sediment plumes.

which allowed for the calculation of light reduction. They were visited weekly. Light reduction was also estimated indirectly from measurements of the growth of discs of the sea lettuce (Ulva), which were placed in Plexiglas cages on buoys in the Øresund. The distribution of the Eelgrass meadows were assessed using aerial photographs taken every autumn.

2.2.4 The mussel programme

The mussel programme also ran every second week, but in contrast to the eelgrass programme, it ran all year round. A device called a photo-sampler was operated from the research vessel. It photographed a fixed area of the seabed and through the use of image analyses the number and size of the mussels on the picture could be measured (see Fig. 2.7).

Every month, additional samples of the mussel beds were taken by divers so that the weight of the live mussels could be assessed. Once a year the distribution of the mussel beds were mapped using the sound analysis software system called RoxAnn. The stations at which sampling took place, by photo sampling or by diver, were chosen on the basis of the sites for ongoing work, baseline studies of mussel bed distribution and hindcast modelling of sedimentation.

The results of the feedback-monitoring were compared with the results of the baseline. If no changes were observed monitoring continued as planned. If a change occurred to the variables chosen in the mussel bed monitoring, the project manager evaluated the reasons for the change. As mentioned under the eelgrass programme, additional sampling may have been necessary also hindcast and forecast sediment modelling was undertaken to assess whether the criteria would be violated. If the project management determined that the change was due to the construction of the Øresund Link and that it would eventually result

Figure 2.7. Photo of a mussel bed in the Øresund.

in a violation of general environmental criteria, feedback action was taken in the same way as previously described.

2.3 RESULTING IMPACT AND CONCLUSION

The construction of the Øresund Fixed Link has been accomplished within the environmental criteria. The impact on measurable variables such as eelgrass and mussels has been as forecasted in the Environmental Impact Assessments. A slight reduction of the eelgrass parameters observed along the west coast of Saltholm indicates that the daily and weekly spill limits were acceptable. During the finalisation of the Flinte Navigation Channel a temporary and permitted exceedance of the weekly spill limits resulted in a local significant reduction of the eelgrass parameters. The eelgrass parameters have now regained their former values due to the spill limitations having been observed.

The final distribution of spilled sediment, which has been calculated by the hindcast simulations, is as predicted in the Impact Assessments. In the future spilled sediment will also be transported around after coming into resuspension during storms, but the intensity and magnitude of this will decrease as more and more of the spilled sediment is deposited in the deeper parts of the Øresund.

The main conclusion of the environmental monitoring programme is that it is possible to comply with very strict environmental regulations if environmental management is incorporated within the organisation.

Chapter 3

Dredging in the Dutch & Belgian Coastal Waters and the North Sea

Prof. Ir. W.J. Vlasblom
*Delft University of Technology, Head Dredging Engineering Section, Delft,
The Netherlands*

3.1 INTRODUCTION

In the coastal areas of North Sea, Rhine-Meuse-Scheldt Delta and Waddenzee are not homogeneous systems, but show great spatial en temporally variation in the morphological en hydro dynamical conditions. A short introduction is given to the geography, hydrography, seabed-topography, seabed-morphology of these areas. The economical activities are in many related to dredging.

3.1.1 Geography

The North Sea located in North-West Europe is surrounded on the west side by Great Britain and on the east side by Norway, Denmark, Germany, the Netherlands, Belgium and France. Its north boundary is mostly set at the 62° North latitude and is linked with the Atlantic Ocean and Norwegian Sea. In the southwest the *Channel* links the North Sea with the Atlantic Ocean as well. To the east it is connected via the *Skagerrak* and the *Kattegat* with the *Baltic Sea*. Including estuaries and fjords, the total surface area is approximately 750,000 km^2 and the total volume 94,000 km^3. This represents less than 1/500 of the total water surface of the earth.

The Netherlands Continental Shelf covers an area of 57.000 square kilometres, roughly 1.5 times the Netherlands. The Belgian continental shelf covers approximately 3,600 km^2, which is about the size of West Flanders. So totally the Dutch and Belgian Continental Shell is less than one per cent of the surface of the North Sea.

A zone of 12 sea miles outside the shore belongs to the territorial waters of the countries, which means that the local law is applicable. These territorial waters and the connected continental shells are a part of the Southern North Sea.

3.1.2 Hydrography

The tides are characterised by a dual periodic variation in the sea level. The tide variations are in general large along the Scottish and English coast, spring tide at least 3 metres and locally more than 6 metres. Along the Dutch and Belgian coast these levels are smaller in general. The tidal regime at the fairway to *Antwerp* reaches up to 125 km upstream with a tidal range between 4.0 and 6.0 m. Tidal range steadily increases in an

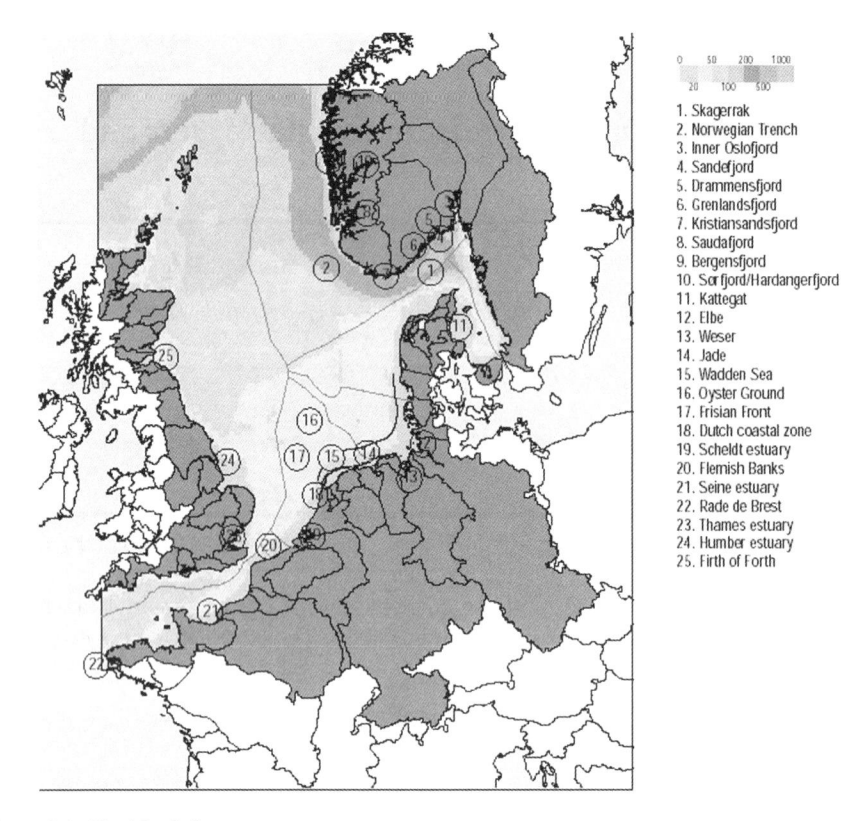

Figure 3.1. The North Sea area.
Source: OSPAR commission 2000. Quality status report 2000, Region II – greater North Sea.

up-estuary direction: at *Hansweert* the mean range is 4.71 m, at *Antwerp* 5.24 m (Reference 2). Flood and ebb durations are asymmetric with the former always of shorter intervals, resulting in flood currents generally be stronger than ebb currents (Reference 3). Near the border between Belgium and the Netherlands peak flood velocities range from 1.33 ms-1 (neap tide) to 2.00 ms-1 (spring tide) and peak ebb velocities are 1.17 ms-1 to 1.67 ms.

The dominant hydrodynamic features of the North Sea are determined by the tide waves from the Atlantic Ocean. The resultant tide current moves from the north along the Scotch and North English coast to the south. A part crosses the North Sea area and flows along the North-Dutch, German, Danish and Norwegian coast to the north. From the south the water flows through the channel along the Belgian and Dutch coast to the north too (Figure 3.2).

The strongest tide currents occur along the coasts; particular along the English coast these currents can exceed values of 1 m/s.

The water depth of the North Sea (Figure 3.3) decreases to the south from more then 200 metre near the *Shetland Isles* and Norwegian coast to 20 metre near Belgian coast. This pattern is broken by the *Doggers Bank* with a depth of 25 metres and a relative deeper part between England and the Netherlands. Along the south-west Norwegian coast lies the 900 km long Norwegian Trench with a depth exceeding the 400 metres (Figure 3.1).

The yearly water temperature differences are large. In the north west of the North Sea the surface temperature varies between 8–13°C and increases in the south east direction to 5–15°C at the *Doggers Bank* and to 4–17°C near the Belgian and Dutch coast.

DREDGING IN THE DUTCH & BELGIAN COASTAL WATERS AND THE NORTH SEA 53

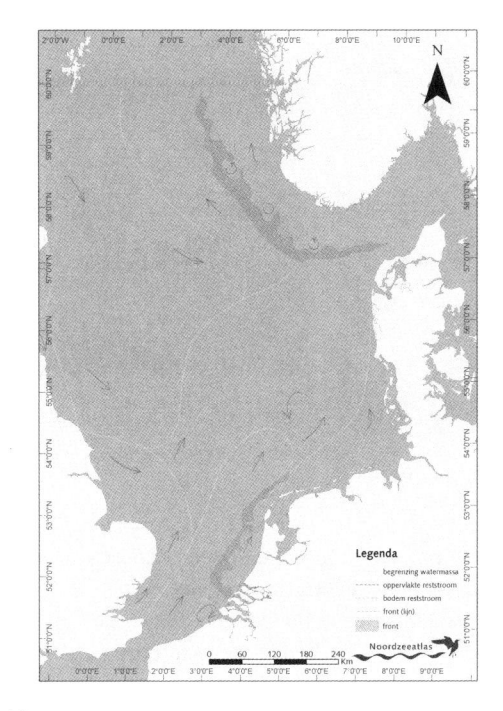

Figure 3.2. Resulting tide currents.

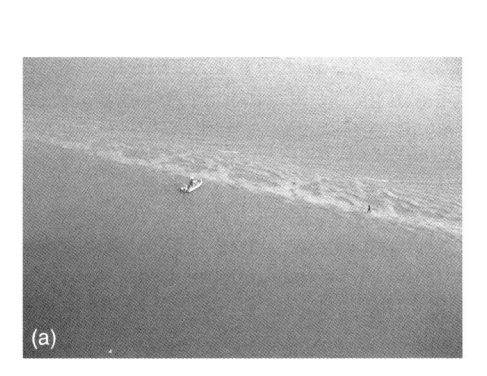

Figure 3.3. Water depth in the North Sea.

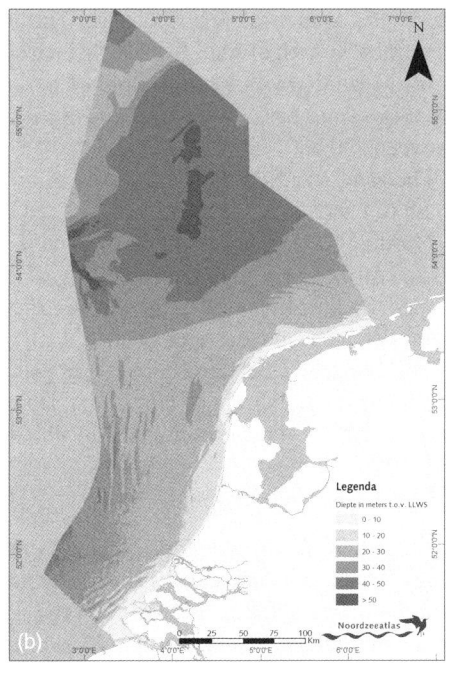

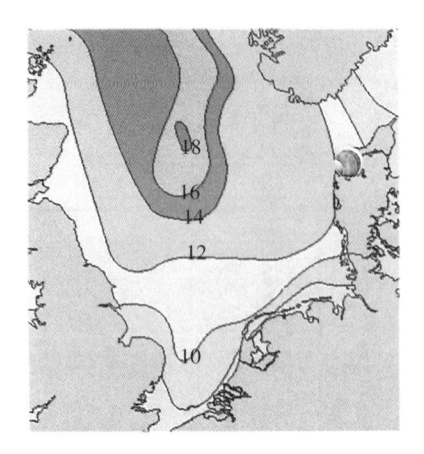

Figure 3.5. Banks emerging from the water.

Figure 3.4. Wave height at the North Sea exceeding once in the 50 years.

Approximately 60 days per year the wind speeds exceeds wind force 7 (storm level), 50% of the time the wave height exceeds the 1.20 m.

Figure 3.4 shows the wave heights at the North Sea, which will be exceeded once in the 50 years. (Source, KNMI). Along the Dutch-Belgian coastal waters these value does not exceed the 8 metres.

3.1.3 Topography

As said the Belgian and Dutch part of the North Sea is shallow. The seabed varies from a flat sandy area to metres high sandbanks. The area along the Belgian coast is characterised by a complex system of sand banks of which some emerges from the water at very low tides.

In the southern part of the Dutch Continental Shelf at depth exceeding the 20 metres are many sand banks with amplitudes between 2 and 12 metre with wavelength varying between 200 and 400 m.

The sand waves in the central part of this area vary in height between 6 metres in the south to 2 metres in the north.

North of the 53° North latitude seabed is mainly flat and the water depth varies between 30 and 50 m. The amplitude of the *Dogger Bank* reaches 25 metres above the surrounding seabed. South-south-east of the *Dogger Bank* a channel, the *Botney Cut*, is present with a water depth varying between 60 and 70 m.

The Dutch and Belgian shore slopes gently from the coastline to a water depth of 15 to 20 metres water. In front of the estuary's in the south of the Netherlands and the *Waddenzee* in the north the slope is dominated by ebb delta's with deep trenches and shallows banks.

3.1.4 Seabed morphology

An important feature which is included in sea bed mapping is that of the sea bed morphology which can vary widely. The Dutch sector of the North Sea contains many different morphological features. In the southern part, where tidal currents are strong, namely the area west of the c. 20 m isobaths and to the north more or less up to 53° N, a huge field of sand waves covers the area with amplitudes varying between 2 and 12 m and wavelength

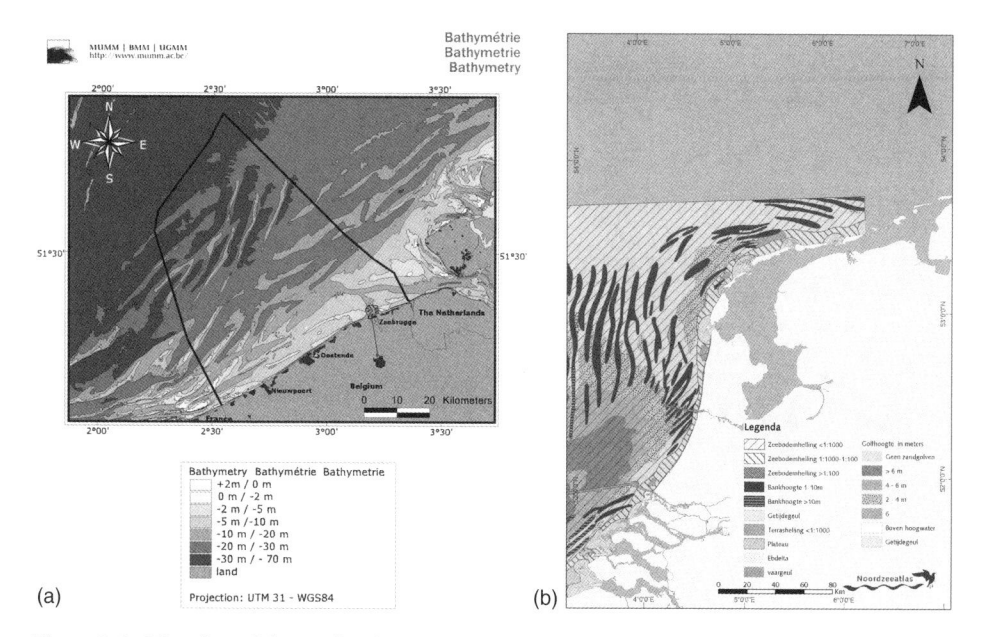

Figure 3.6. Direction of the sandbanks along the Belgium and Dutch coast.

varying between c. 200 to c. 400 m. The amplitude decreases in height to the north. To the south-west along the coast as far north as 52° N huge linear sand banks, the Zeeland Ridges, occur with heights of >20 m and reach to just below sea-level. In the area north of c. 52°30′ N a pattern of more or less north south running linear ridges with an amplitude of <10 m and length up to >100 km is present. The amplitude of one of these banks, the Brown Bank, reaches to 19 m below sea level.

North of c. 53° N the seabed is mainly flat and the water depth varies between 30 and 50 m. The eastern part of the *Dogger Bank* occurs to the north-west with an amplitude of 25 m above the surrounding sea bed. South-south-east of the *Dogger Bank* a channel, the *Botney Cut*, is present with a water depth varying between 60 and 70 m.

An interesting study is done by Van Lancker (2000), e.a about sediment transport pathways in the nearshore area of the western Belgian coastal zone. The area is characterised by a bank–swale hydrography witnessing water depths of 0 to −15 m MLLWS (Figure 3.7).

The transport of sediments and other marine properties in the coastal zone is strongly affected by a variety of hydrodynamic and meteorological processes. Especially in shallow waters, wind induced residual currents can not be neglected, moreover the wave climate has a direct effect on sediment suspension and bedform morphology. The area under investigation is characterised by semi-diurnal tides. The tidal currents are asymmetrical with a north easterly-directed residual flood displacement that controls the sediment transport pathways. In the Westdiep (Figure 3.8) swale, spring near-surface flood currents in a NE-ENE direction can amount up to 1.32 m/s whilst a maximum of 0.86 m/s is reached during the SW-WSW ebb currents. At this location, the tidal amplitudes at spring and neap tides are 5.42 m and 2.89 m respectively.

Generally, the area is characterised as having a high potential for the deposition of fine-grained sediments. The sedimentological investigation of the present study has shown

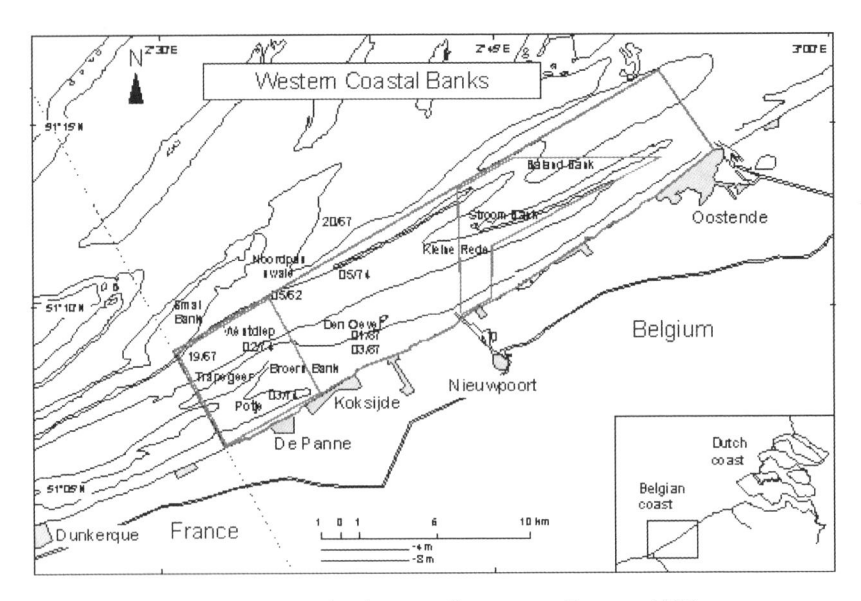

Figure 3.7. Localisation of the Broers bank – Westdiep system (Degraer 1999).

that the sand and silt fractions are the main constituents of the surficial sediments within the Broers Bank – Westdiep system. The sediments get coarser towards the top of the bathymetric highs, including a rise in the Westdiep swale (thick contour line of −13 m MLLWS, Figure 3.8). The sand bank areas are indeed generally coarsest, whilst the surficial sediments of the swales can have high percentages of the silt fraction. The shallow Broers Bank (up to 0 m MLLWS) is characterised by coarse shell fragments. Still, the sorting is, generally, best on the bank summits and poorest in the swales. Throughout the study area, the mean grain-size ranges from fine sands of 2.75 phi (∼150 m) up to medium sand of 1 phi (500 m). Figure 3.8 is a contour map of the distribution of the fine and medium sands. From the side-scan sonar registrations it appeared that the larger sand dunes have a strike perpendicular to the maximum tidal current velocity (Figure 3.9).

3.1.5 Seabed sediments

The North Sea area is covered by huge sediments layers deposit in the Quaternary South of 53° N of the Belgian-Dutch Continental Shelf (Figure 3.10).

The superficial sediments consist mainly of fine to medium sand, which consist of reworked fluviatile deposits of the rivers *Rhine and Meuse*. North of 53° N the seabed sediments become finer and consist of mainly reworked glacial deposits. In the area north of 54° N a huge area with a mud percentage of between 10–90% is present. In this area the tidal currents are very low because of the existence of an amphidromic point. Because of decreased tidal currents the mud drops out of the water column and deposits. Near the coast much silt can be expected.

3.1.6 Economy

In the last century the North Sea and its coastal areas has become an important engine for the Belgium-Dutch economy. Many human activities are taken place nowadays.

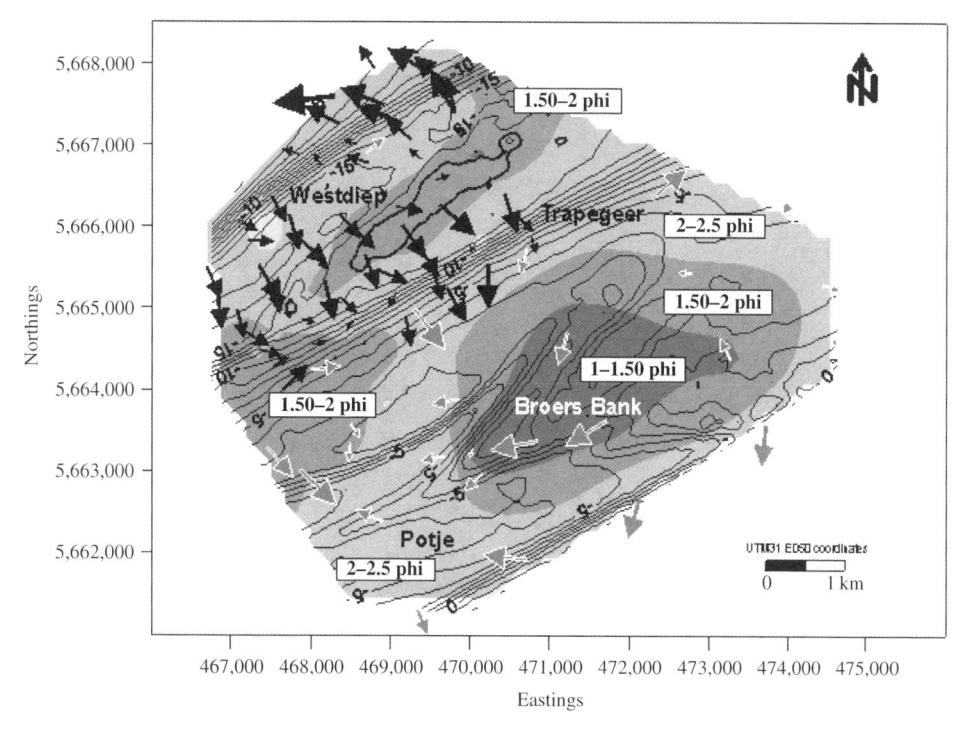

Figure 3.8. Contour map of the mean grain-size of the surficial sediments (moment values). (medium sand: 1–2 phi/250–500 µm; fine sand: 2–3 phi/125–250 µm) (Honeybun, 1999). The vectors represent residual transport directions based on a grain-size trend analysis according to Gao & Collins (1992).

The Progress Report of the Fifth North Sea conference in 2002 in Bergen mentions explicitly the following human activities (pp 23):

- The North Sea is one of the world's most important areas for harvesting fish and shellfish.
- Coastal industries of various types are located along the coasts and estuaries of the North Sea, discharging pollutants to marine waters and in some instances requiring large amounts of cooling water.
- The North Sea contains some of the busiest shipping routes in the world, and most of Europe's largest ports are situated on North Sea coasts and rivers.
- The offshore oil and gas industry has become a major economic activity in the North Sea since the late 1960s. There are about 130 oil and gas production platforms on the Dutch Continental Shelf. About 8% of these platforms are for the extraction of oil, the remainder for gas.
- Mariculture for fish and shellfish is undertaken in many of the North Sea States.
- Coastal engineering includes damming of rivers, but also beach nourishment, diking and land reclamation.
- Tourism in North Sea coastal areas and adjacent land is an important social and economic activity with intense development pressure.
- Mineral extraction (sand, gravel and rocks, calcium carbonate (shell aggregates, marl)) takes place in many North Sea States.

58 DREDGING IN COASTAL WATERS

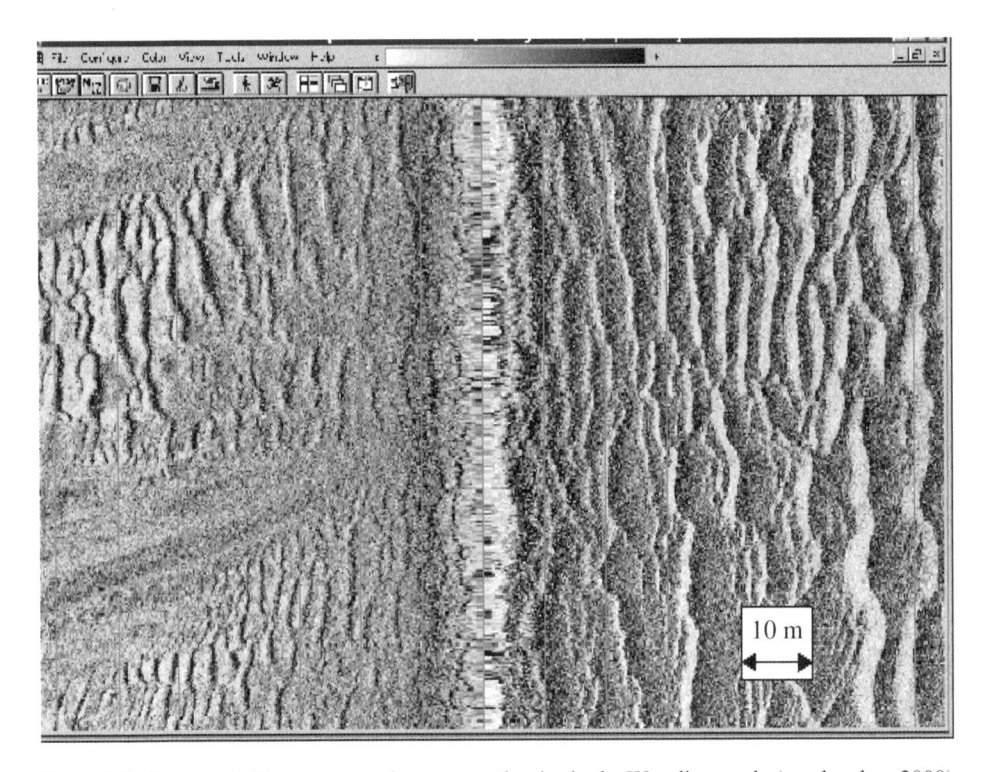

Figure 3.9. Corrected side-scan sonar image near the rise in the Westdiep swale (van lancker, 2000).

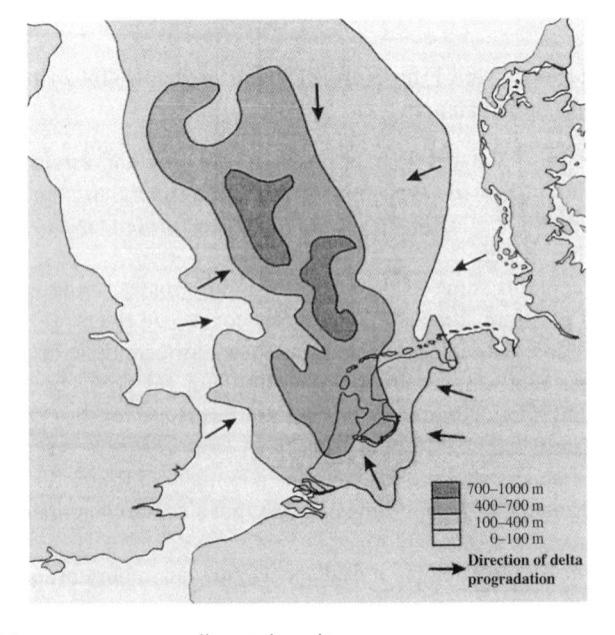

Figure 3.10. Thickness quaternary sediment deposits.

- Dumping of dredged material (for maintenance dredging, laying of cables and pipelines), waste from fish processing and inert material of natural origin.
- Power generation by tidal or wave energy is limited to a few potential locations, but offshore windmills will increase in number. The Netherlands goal for the development of wind energy is 6,000 MW wind energy from the Exclusive Economical Zone in 2020.
- Military uses of the sea in peacetime include fishery protection patrols and NATO exercises.

Many of those activities are related to dredging in a certain way.

3.2 DREDGING ACTIVITIES

The Netherlands yearly borrows 31 Mm^3 sand from the North Sea, from which 10 Mm^3 is dredged in the fairways of Rotterdam and Amsterdam to guarantee the required depth. 15 m^3 is necessary for beach nourishment and fore shore supply. From the resulting 16 Mm^3 94% is used for inland reclamations and 6% is used for industrial purposes.

3.2.1 General

The dredging activities executed in the Belgian-Dutch coastal waters have different goals:

1. Maintaining the water depth in fairways, (approach) channels and harbours.
2. Port development of harbours (*Rotterdam. Amsterdam, Antwerp, Zeebrugge*, etc)
3. Protection against the sea (beach nourishment, diking)
4. Reclaiming of coastal areas
5. Borrowing of sand for inland reclaiming purposes.
6. Sand, gravel and shells extraction for industrial purposes.
7. Burying of pipelines and cables.

Combination of these activities is usual to reduce cost and environmental impact. Permits provided by governments control these activities.

3.2.2 Maintenance dredging

Dredging to be carried out to maintain the maritime access routes to the Belgian and Dutch coastal ports and the depth of these ports are the responsibility of the Flemish Region and the Dutch ministry of Transport. The large quantities of dredged material resulting from these activities, which may be polluted to varying degrees, are either dumped back in the sea, or store in basins. This procedure is the responsibility of the federal environment department. It can have an impact on the marine ecosystem.

As said, the North Sea is rather shallow and if nature had his way, the fairways to *Zeebrugge*, *Rotterdam* and *Amsterdam* should have a depth of maximum 10 metres. Maintenance dredging in the fairways to these harbours is almost a continuous job.

The fairway to *IJmuiden* and Amsterdam, called *IJ-geul* is maintained on a depth of 19 m below mean sea level. Vessels with a maximum draught of 16.5 m may enter these harbours.

The fairway to Rotterdam "*Eurogeul*" is maintained 24 metre and is therefore suitable for ships with a maximum draught of 22,5 metre.

The fairway to *Zeebrugge* and to *Antwerp* has a maintained depth of 13.5 m LLW.

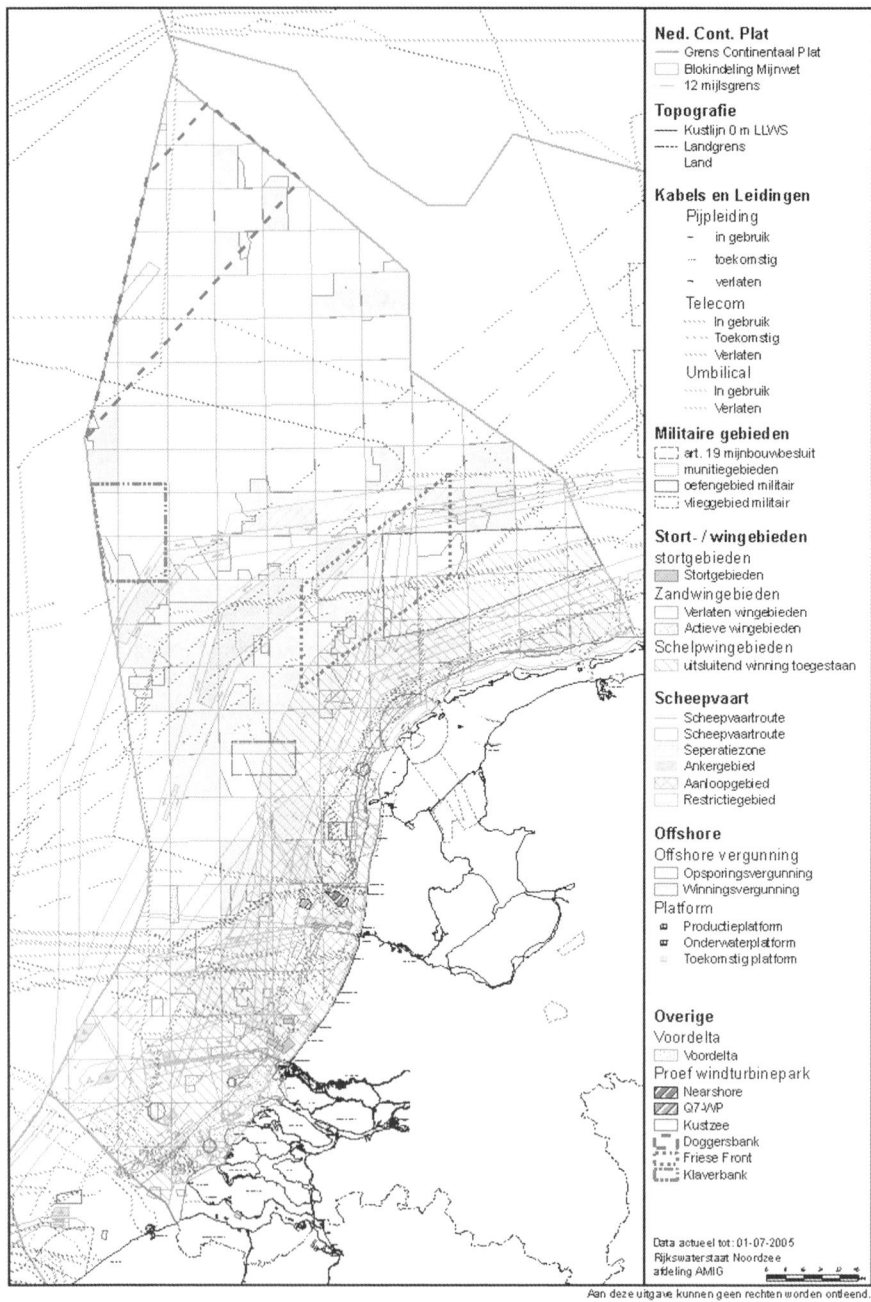

Figure 3.11. Activities at the North Sea (Source: RWS).

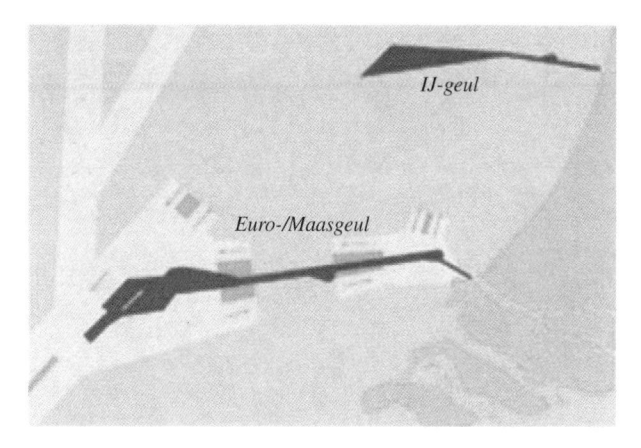

Figure 3.12. Euro-/Maasgeul and IJ-geul; the fairways to Rotterdam and Amsterdam.

Nowadays ships with draughts of up to 15.39 m can reach the port of *Antwerp* on a single tide. Draughts of up to 15.54 m can reach the port on two tides.

Maintenance dredging is executed mainly by Trailing Suction Hopper Dredgers (TSHD's). The main reason is that this self-propelled dredger needs no anchors and wires but sails freely while dredging in the approach channels and harbours. The dredged material in the coastal waters and harbours is silt and/or sand.

Sand and Silt can be brought into harbour by the tidal currents or by the rivers *Rhine*, *Meuse*, and *Scheldt*.

The silt is either stored in basins when polluted (mostly from inland sources) or dumped at sea at the designated dumping areas. The sand is used for beach nourishment and fore shore protection or partly brought to shore where it is desalted and used for reclaiming purposes.

3.2.3 Port extensions

Today, port extensions are strongly related to the increase in container transport. Many ports have plans for extensions of their container facilities; *Rotterdam* their second *Maasvlakte*, *Antwerp*, *Zeebrugge*. Besides extensions of the existing container handling facilities, a number of non container ports have plans for container terminals too (*Vlissingen* Container Terminal *Westerschelde*, *Delfzijl*, *Eemshaven*). On the other hand the container terminal of the Port of *Amsterdam* is still idle for more than two years after completion.

The type of dredging equipment used for this type of work, depends geographic situation; for inland port extension Cutter Suction Dredgers and/or Plain Suction Dredgers are used. Offshore port extensions mainly by Trailing Suction Hopper Dredgers.

3.2.4 Beach nourishment

The erosion of the Dutch-Belgium North Sea coast is in certain area that high that additional coastal protection is necessary. (Examples: *De Haan*, Belgium Figure 3.12; *Egmond* and *Callandsoog*, The Netherlands.)

Mostly, this is done by beach nourishment or foreshore supply. Sand from designated borrow-areas is either pumped on the beach by means of the so-called pump ashore installation or dumped as an embankment parallel with and as close as possible to the

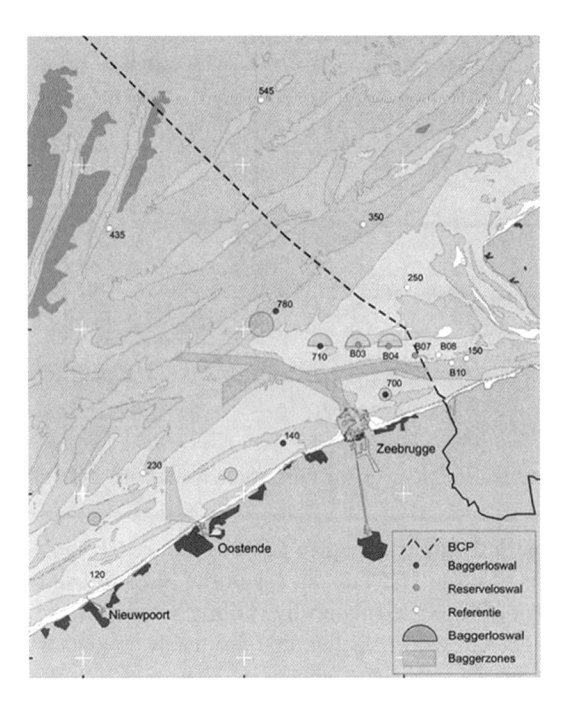

Figure 3.13. Fairways to Zeebrugge and Antwerp.

beach. Particular for this last type of work small shallow draft Trailing Suction Hopper Dredgers (TSHD's) are used. When pumping distances to the shore becomes to large for the TSHD the sand can be dumped in a "rehandling" pit from which a cutter suction dredger will pump it to the shore. A cutter suction dredger (CSD) has almost the full cycle time of the TSHD available to pump the material to the shore. So its mean production is much lower than when the load is pumped to the shore by the TSHD and can therefore pump further with the same power.

Such a situation is comparable with the picture besides. A CSD working as a suction dredger (without a cutter) borrows sand that she pumps via booster to the *Groninger Sea Dike* (Figure 3.16).

3.2.5 Dredging activities for pipelines, outfalls & shore connections

Besides dredging for coastal protection dredging is done for burying of pipelines and cables at sea and to the shore. Figure 3.14 shows a cutter dredger dredging a trench for burying the connection of sea gas pipeline to the shore gas net. Burring cables is done with special vessel and mostly during laying. Pipelines either are lay down in trenches dredged by TSHD's or on the seabed and buried by post trenching techniques with special equipment.

3.2.6 Reclaiming coastal area's

In the past a number of coastal areas have been reclaimed from the sea; the *Port of Zeebrugge*, "*the Slufter*", a depot for contaminated dredged material, *van Dikshoorn triangle*, an area reclaimed connected to the North Pier of the entrance to the harbour

Figure 3.14. A mud flow entering the harbour of Zeebrugge.

of *Rotterdam, Marina IJmuiden*, an area connected to the South Pier of the entrance to the harbour of *Amsterdam*.

The dikes of *the Slufter* are built from sand dredged from the interior of *the Slufter* by Cutter suction dredgers and plain suction dredgers. The *van Dikshoorn triangle* and *Marina IJmuiden* are reclaimed by using trailing suction hopper dredgers borrowing the sand from the fairways to the respectively harbours.

Further port extensions (Rotterdam the new container terminal "*Maasvlakte II*"; *Container Port Vlissingen* and at the Port of *Zeebrugge*) can expected in the near future.

Besides the creation of new land, large scale extraction of sea-sand will be required in response of the expected increase in erosion of beaches and dunes.

In the last 35 years many studies has been made for reclaiming large coastal areas or islands at sea for different purposes; living areas along the coast, industrial islands, airport at sea, a main port combined with an airport, but till today none of these plans are realised due to the cost involved and the uncertainty regarding the ecological and morphological effects.

One of the ecological questions for large scale sand extraction from the coastal areas is, whether extraction should result in a small but deep pit or in a large shallow area. The first has the advantage that the area of benthic live is small compared by the other solution, however deep sand pits will have a significant influence on the flow and wave pattern, which may result that the pit acts as a sink for sediments from the nearby areas, which can on its turn lead to the erosion of the nearby seafloor.

3.2.7 Sand & gravel extraction

North Sea sediments mainly consist of fine to medium sand. Coarse sand and gravel exist mostly in relatively small patches at some depth below the seafloor.

Figure 3.15. Beach nourishment at DE HAAN, Belgium.

Figure 3.16. Dredging in the Waddenzee. Figure 3.17. Shore connection gas pipeline.

These aggregates, encountered on the Belgium Dutch part of the North Sea, are already been successfully dredging for decades by England, Belgium and the Netherlands. Borrowing is permitted in the designated areas. Over the past few years, a steadily growing interest in the use of this sea sand has been observed. This interest has grown out of the depletion of existing sand quarries on lad, the alternative use of these often beautiful regions as sites for new residential areas, etc. and the growing demand for sand and gravel for industrial purposes.

Sea sand is used for three specific purposes: in the construction sector, which accounts for approximately one tenth of Belgium's total sand production, as beach supplements, to curb the erosion of the Belgian coast as a result of currents, waves, etc. and for land reclamation which, unlike in the Netherlands, is undertaken exceptionally in Belgium. The import of sea sand in the Netherlands comes partly from maintenance dredging of the

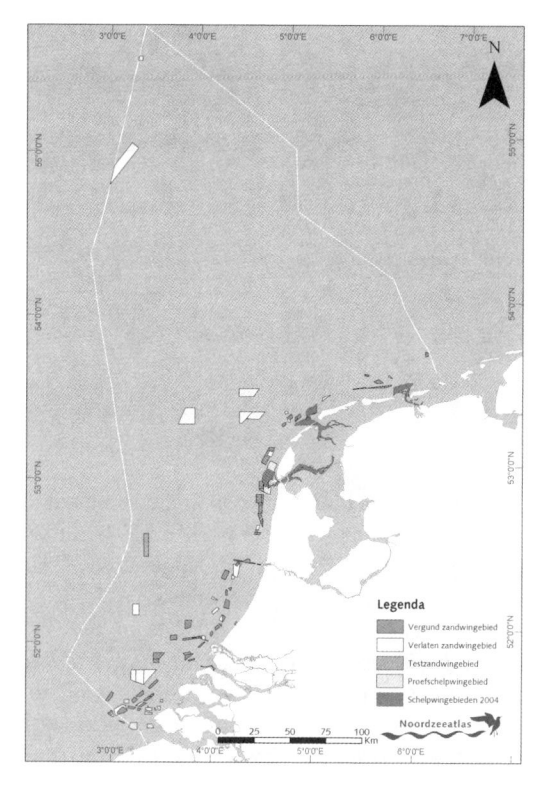

Figure 3.18. Sand, gravel- en shell borrowing.

fairways to the ports. The yearly quantity of sea-sand is about 16 million m^3 (2002), in Belgium this quantity is much lower, about 2 Mm^3. There is no gravel extraction on the Dutch Continental Shell.

To reduce the effects from these activities on the coastal protection and the ecology in the coastal zone dredging is only permitted in the fairways and outside the -20 m NAP bathymetry line till a depth of 2 m below seabed. An exception is made for shells; dredging outside the -5 NAP bath line.

The dredging is mainly done by trailing suction hopper dredgers and controlled by government with black boxes.

3.2.7.1 *Shell extraction*

On basis of the National Policy Plan the total permissible amount (in m^3) of shells to be extracted from the Wadden Sea, North Sea and Internal Waters is restricted to 210,000 ton per year for the Netherlands.

3.2.7.2 *Environmental consideration about sand and gravel extraction*

The influence of sand and gravel extraction on the animal population in the area of extraction was shown at an experimental site of 200 m × 500 m, which was dredged over a four day period in April 1992 by a trailing suction hopper dredger. The sediment type at the test site was similar to commercial gravel beds and was considered to be representative of an undisturbed seabed. About 70 per cent site was dredged and 50,000 tons of material

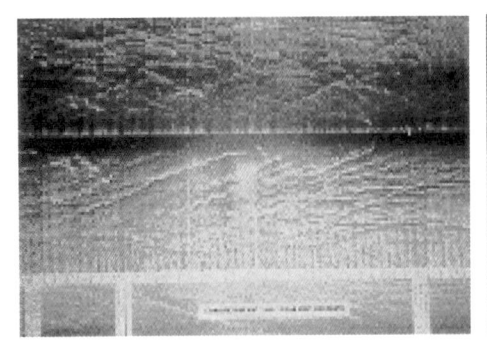

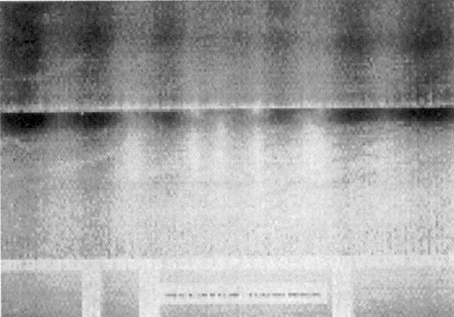

Figure 3.19. A side-scan sonar trace showing the dredge tracks two weeks after completion of dredging.

Figure 3.20. A side-scan sonar trace showing the dredge tracks two years after completion of dredging.

removed. The dredging caused immediate changes to the sediment and the animals living in and on the sediment. The sediments within dredged areas became finer and much more mobile as the dredging tracks were infilled and her ridges eroded. The animal populations were reduced much in abundance, diversity and biomass.

Re-colonisation by some species was fairly rapid. During the first 12 months after the dredging was stopped nearly 70 per cent of the original types of animals present was re-established but abundance and biomass remained significantly depressed. The diversity of animals in samples were not significantly different from those at the nearby undredged reference site, but the abundance and biomass were stilt significantly lower. A reduced biomass was expected since newly established colonies are likely to be smaller in size and weight than mature colonies at undisturbed sites.

Surveys in 1995 showed that both the diversity and biomass of animals were indistinguishable from those on the reference site and matched the colonies on the experimental site before dredging. The abundances at the dredged site appear to have stabilized, at a level significantly below those at the reference site.

Two years after dredging, a side-scan survey showed that the dredged tracks had almost completely disappeared (Figures 3.19 & 3.20).

3.3 DREDGING EQUIPMENT

The size of the dredgers depends on the required productions e.g. volumes dredged per unit of time, the size and depth of the harbours and fairways and protected or unprotected waters.

Dredging equipment can be divided in hydraulic dredgers and mechanical dredgers (Figures 3.21–3.26), named to the way the dredger transports the material from the seabed above water level.

With the exception of the Trailing Suction Hopper Dredger, all dredgers are unsuitable for working under offshore conditions, besides they are special designed for it. However, when the sea state is quiet they can work offshore of course, but shelter areas should be close by in the case the sea state changes.

The excavation of the soil can be hydraulically or mechanically too. Hydraulic excavation is done in cohesionless soils, such as silt sand and gravel, by the suction flow or water jets.

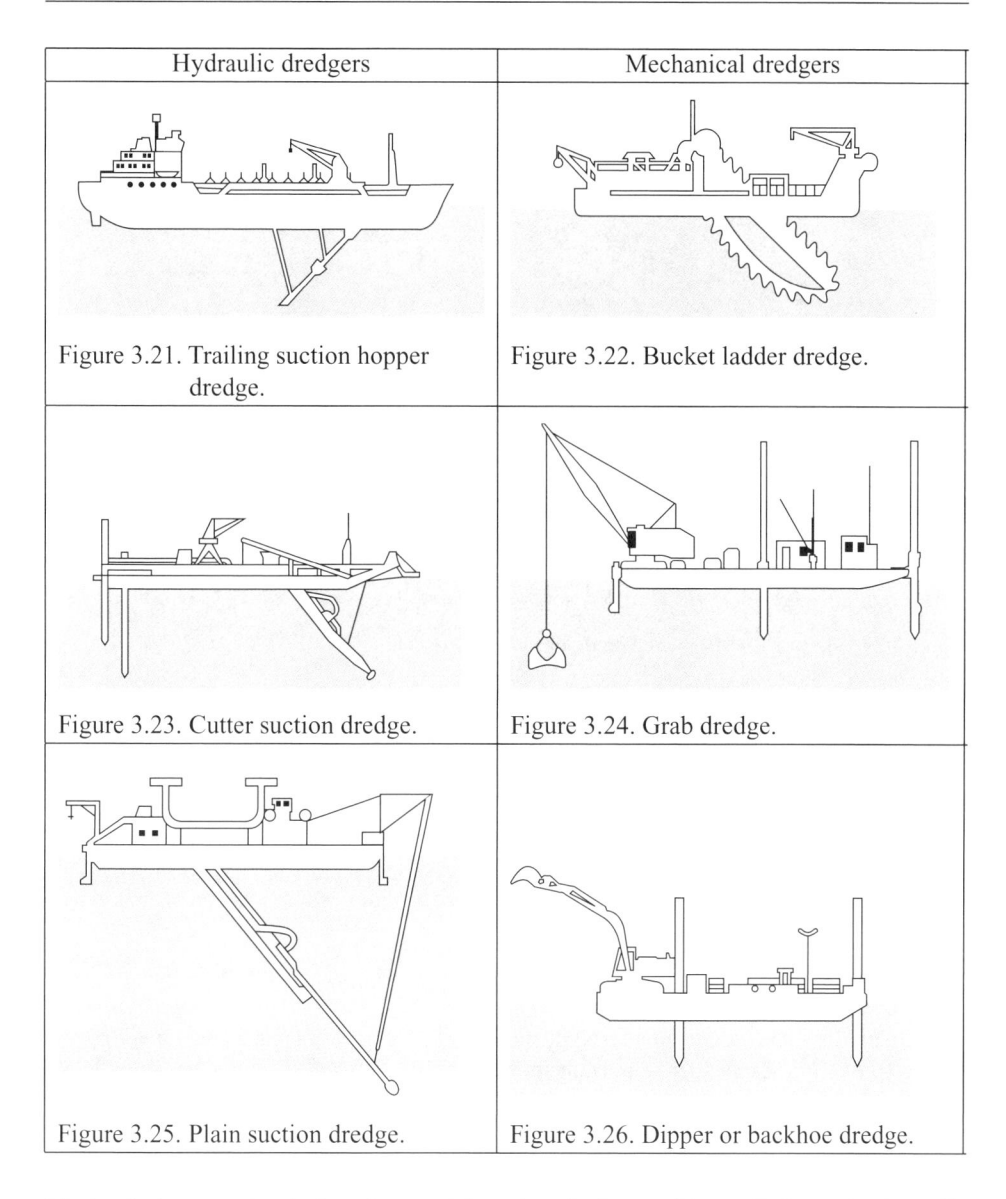

Hydraulic dredgers	Mechanical dredgers
Figure 3.21. Trailing suction hopper dredge.	Figure 3.22. Bucket ladder dredge.
Figure 3.23. Cutter suction dredge.	Figure 3.24. Grab dredge.
Figure 3.25. Plain suction dredge.	Figure 3.26. Dipper or backhoe dredge.

Mechanical excavation is done by special cutting devices and is applied to cohesive soils and sand.

3.3.1 The trailing suction hopper dredger

The characteristics of the trailing suction hopper dredger are that it is a self-propelled sea or inland waterway vessel, equipped with a hold (hopper) and a dredge installation to load and unload itself. In a standard design the trailing suction hopper dredger is equipped with:

- One or more suction pipes with suction mouths, called dragheads that are dragged over the seabed while dredging.
- One or more dredge pumps to suck up the loosened soil by the dragheads.

Figure 3.27. View on a TSHD while dredging.

- A hold (hopper) in which the material sucked up is dumped.
- An overflow system to discharge the redundant water.
- Closable doors or valves in the hold to unload the cargo.
- Suction pipe gantries to hoist the suction pipes on board.
- An installation, called the swell compensator, to compensate for the vertical movement of the ship in relation with the seabed.
- Nowadays most of these ships are equipped with a pump ashore system, which makes possible to pump the material ashore.

The main advantages of a trailing suction hopper dredger are:

- The ship does not dredge on a fixed position. It has no anchors and cables, but it moves freely, which is especially important in harbour areas, where it is not an obstacle for shipping.
- The trailing suction hopper dredger is quite able to work under offshore conditions.

3.3.1.1 *Applied working area*

The TSHD is a free sailing vessel and does not hinder other shipping during dredging and is therefore ideal for dredging in harbours and shipping channels inshore as well as offshore. The seagoing vessels are very suitable for borrowing sand under offshore conditions (wind and waves) and large sailing distances. The dredged material is dredged, transported and discharged by the vessel without any help from other equipment.

(De)mobilization is very easy for this type of dredger. It can sail under its own power to every place in the world under the condition of sufficient water depth.

Suitable materials for the TSHD to dredge are soft clays, silt sand and gravel. Firm and stiff clays are also possible but can give either blocking problem in the draghead and/or track forming in the clay. In that case the draghead slips into foregoing tracks, resulting in a

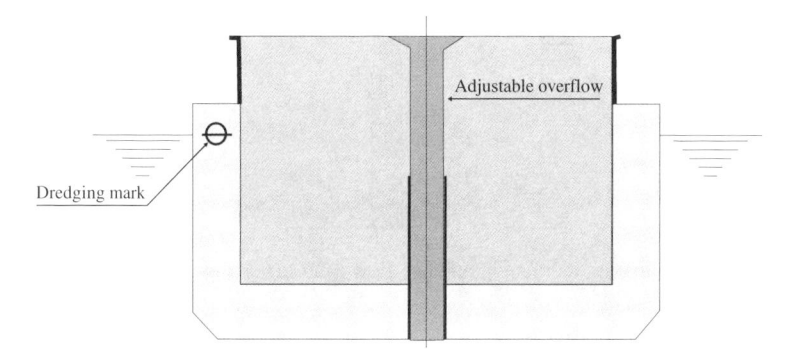

Figure 3.28. Adjustable overflow.

very irregular dredged surface. Dredging rock with a TSHD is in most cases not profitable. It requires very heavy dragheads with rippers while the productions are rather low.

The trailing suction hopper dredger sucks the soil from the seabed at a sailing speed of 1 to 1.5 m/s (2 to 3 knots) and deposits it in the load called hopper. For non- or bad-settling soils the dredging is stopped when the surface of the mixture in the hopper reaches the upper edge of the overflow (Figure 3.28).

When dredging settling soils, the dredging will be continued after the mixture reaches the maximum level of the overflow. Because, most of the solids settle before it reach the overflow. The remainder, the fine material, is discharged with the water through the overflow.

The dredging is stopped when:

- The hopper is full with a mixture and overflowing is not allowed, for instance when dredging (contaminated) silt (Figure 3.28).
- The maximum allowable draught is reached and the overflow can not be lowered usefully anymore (Figure 3.30).
- The economical filling rate is reached.

When dredging stops, the suction pipes are pumped clean to prevent settling of the sand or gravel during the hoisting of the pipes causing an extra load for the winches. When the

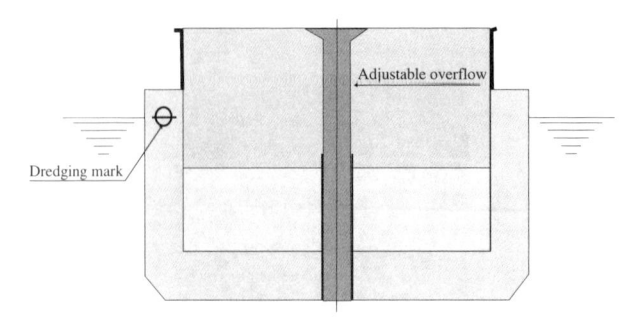

Figure 3.29. Overflow at maximum level and TSHD at maximum draught when dredging settling slurries.

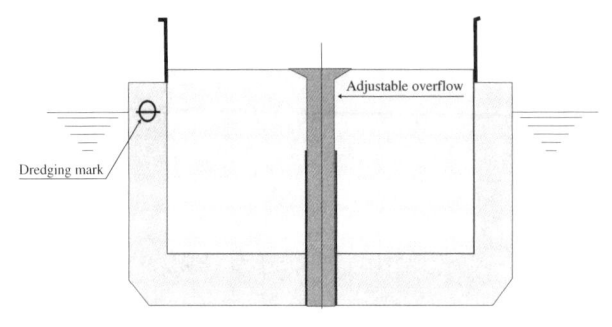

Figure 3.30. TSHD at maximum draught and overflow can't lower anymore when dredging settling slurries.

pipes are cleaned the pumping stops and the pipes are raised. When the dragheads are out of the water the ships velocity is increased to sail to the discharge area.

The discharge area can:

- Be in its most simple shape a natural deepening of the seabed, called the dumping area (or shortly the dump), to store redundant material. If the storage capacity is large, there is no concern about the way of dumping. This hardly happens nowadays. The client demands usually a dump plan to fill the dump area as efficiently as possible. At all times the draught on the dump needs to be sufficient to open the bottom doors or valves (Figure 3.31).
- Be a storage location for contaminated silt, like for instance the Slufter (Rotterdam harbour). Here the material is pumped ashore using a pump ashore discharge system (Figure 3.32 and Figure 3.33).
- An area that has to be reclaimed.
- An oil or gas pipe that has to be covered.

In case of the discharge area is a **dump**, opening the doors or valves in the base of the hopper does the unloading.

This is usually done with an almost non-moving ship, certainly when accurate dumping is required. During the dumping water is pumped onto the load by means of the sand pumps. The eroding water stimulates the dumping process. If the trailing suction hopper dredger is equipped with **jet pumps** connected to a jet nozzle system in the hopper, those will be used too. The jets more or less fluidize the load and improve the dumping process.

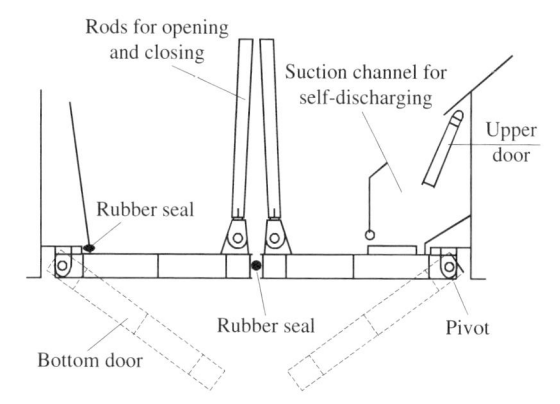

Figure 3.31. Bottoms doors operated by rods.

Figure 3.32. Pump ashore connection.

Figure 3.33. TSHD Poseidon picking up the floating pipeline to pump her load to the shore c.

The **shore connection**, being the connection between the board pipeline system and the shore pipeline is currently mostly positioned just above the bow (Figure 3.32). The connection between the ship and the shore piping is this case a rubber pipeline. (Figure 3.33) The ship remains in position by manoeuvring with its main propellers and bow thruster(s).

When pumping the ashore by using the sand pumps the aforementioned jets in the hopper are available to fluidize or erode the load. After the load is either dumped or pumped ashore, the ship will return to its suction area and a new cycle starts. In general the ship sails empty, in a non-ballast way, back to its suction section. There is only some residual water and/or load left in the hopper.

3.3.1.2 *Environmental considerations*
This type of dredger can induce turbidity in two ways:

1. When dredging the suction head is dragged over the seabed with a speed of 1 to 1.5 m/s. The dragging of the heavy device introduces a turbidity cloud around the draghead, which settle mostly to the background turbidity when dredging is stopped.
2. During dredging the dredged sand-water mixture is dumped into the hopper. Most of the sand particles will settle and the remainder is discharged with the water through the overflow, inducing turbidity in the water column.

Particular to the last point recent years new research has initiated by the dredging industry.

Figure 3.34. Reclamation area.

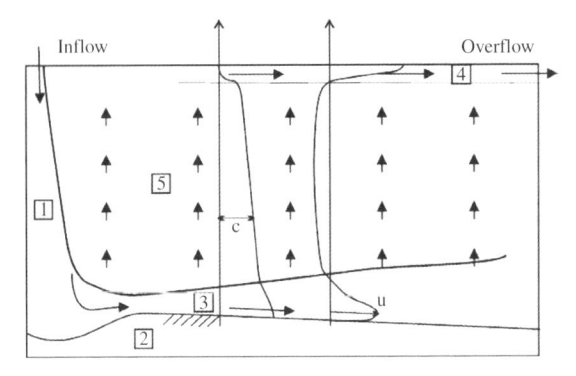

Figure 3.35. Flow and concentration during loading.

According to van Rhee (2002), the flow pattern in the hopper during the overflow phase can be schematised as indicated in Figure 3.35. The inflow is located at the left side and the overflow at the right side of the hopper.

The hopper area can be divided into 5 different sections:
1. Inflow section
2. Settled sand or stationary sand bed
3. Density current over settled bed
4. Horizontal flow at surface towards the overflow
5. Suspension in remaining area

A measure for the quality of the settling process is the relative cumulative overflow loss. This is defined as the ratio between the total amount of solids that leave the hopper through the overflow and the total amount of solids pumped in the hopper

$$OV_{cum} = \frac{\int_0^T c_o(t)Q_o(t)dt}{\int_0^T c_i(t)Q_i(t)dt}$$

In which c is the concentration and Q the discharge. The indices o and i relate to the overflow and inflow respectively.

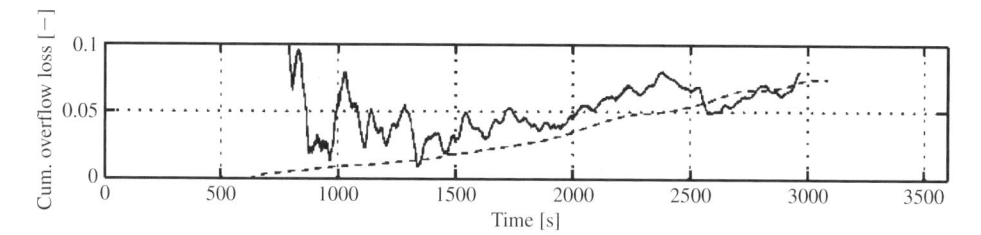

Figure 3.36. Model test results of overflow losses.

This relative cumulative overflow loss is, except for the material properties as grain size, the grain distribution, shape and specific mass, also dependent on the loading conditions like the flow rate, concentration, turbulence intensity, temperature and the hopper geometry.

Figure 3.36 below shows the measured and calculated cumulative overflow losses on board of the TSHD Cornelia during a trip (van Rhee, 2002). The cumulative losses increases with time.

In his thesis, van Rhee defined a dimensionless overflow rate as being the ratio of the settling mass flux over the inflow mass flux.

$$S^* = \frac{S_{in}}{S_{sed}} = \frac{c_{in}}{c_b} \frac{1-n_0-c_b}{1-n_0} \frac{Q}{BLw_s} = \frac{c_{in}}{c_b} \frac{1-n_0-c_b}{1-n_0} \frac{Q}{w_0(1-c)^n BL}$$

$$= \frac{c_{in}}{c_b} \frac{1-n_0-c_b}{1-n_0} H^*$$

In which:

Symbol		Description	Dimension
S_{in}	=	Inflow sand flux	[t/s]
S_{sed}	=	Sedimentation flux	[t/s]
c_{in}	=	Concentration at inflow	[–]
c_{bed}	=	Concentration sedimentation flux	[–]
n_0	=	Porosity	[–]
Q	=	Inflow	[m³/s]
B	=	Hopper width	[m]
L	=	Hopper length	[m]
w_s	=	Settling velocity particles	[m/s]
w_0	=	Non-hindered settling velocity particles	[m/s]
P_s	=	Particle density	[kg/m³]

And $H^* = \dfrac{Q}{w_0(1-c)^n BL}$ called the dimensionless hopper load

Further is:

The inflowing massflux equals: $S_{in} = \rho_s Q c_{in}$ Where c_{in} is the inflow concentration.
The massflux settling in the bed becomes: $S_{sed} = \rho_s(1-n_0)BL v_{sed}$.

A relationship between the overflow losses and the dimensionless overflow rate is expected with the measured cumulative overflow losses at the end of a test (fully loaded hopper). However the near bed concentration c_b is not known a priori. As a first approximation $c_b = c_{in}$ can be used.

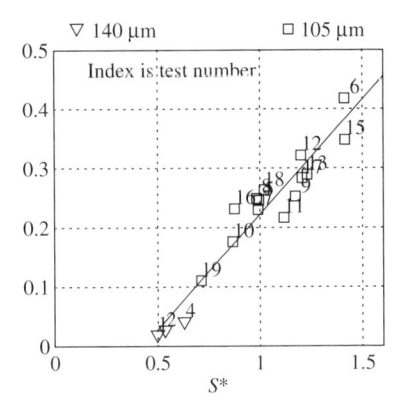

Figure 3.37. Cumulative overflow losses versus dimensionless overflow rate.

Figure 3.38. The cutter suction dredger.

The cumulative overflow losses from the model test are plotted versus S^* (so without the influence of erosion) in Figure 3.36. This figure shows clearly a reasonable relationship between losses and dimensionless overflow rate S^*.

The best fit through the data points can be expressed as: $OV_{cum} = 0.39(S^* - 0.43)$

The following influences are not included in the value of S^*:

- The loading and overflow structure.
- The horizontal transport of sediment (density current), which reduces the sedimentation and may cause scour.
- The shape of the grain size distribution.

Nevertheless the overflow loss is reasonable predicted by this equation.

General Considerations

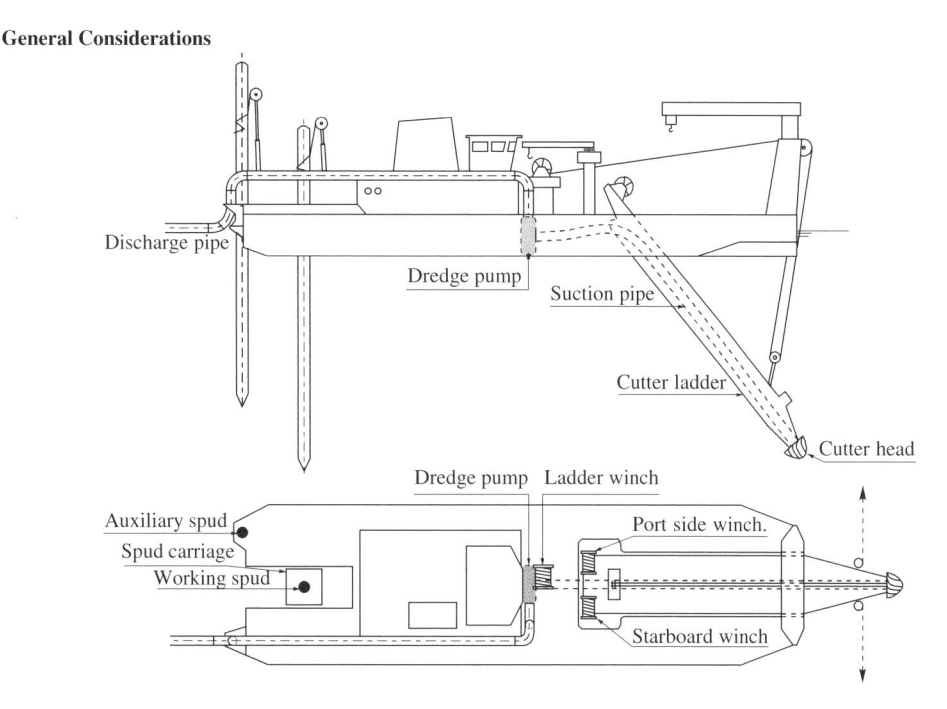

Figure 3.39. Lay-out of cutter suction dredger.

This indicates that these influences did not play a major role during the laboratory experiments. However, prototype tests have showed that some of the influences can affect overflow losses. The simple equation above must therefore be regarded as a lower limit for the overflow losses.

When dredging silt the overflow losses becomes that high that the dredging is stopped when the overflow reached. This may be either by economical reasons or by environmental reasons.

3.3.2 The cutter suction dredger

3.3.2.1 *General considerations*

The cutter suction dredger is a stationary dredger equipped with a cutter device (cutter head), which excavates the soil before it is sucked up by the flow of the dredge pump(s).

During operation the dredger moves around a spud pole (Figure 3.40) by pulling and slacking on the two fore sideline wires. This type of dredger is capable to dredge all kind of material and is accurate due to their movement around the spud pole. The stationary cutter suction dredger is to distinguished easily from the plain suction dredger by its spud poles, which the last don't have.

The ladder, the construction upon the cutter head, cutter drive and the suction pipe are mounted, is suspended by the pontoon and the ladder gantry wire.

The spoil is mostly hydraulically transported via pipeline, but some dredgers do have barge-loading facilities as well (Figure 3.42).

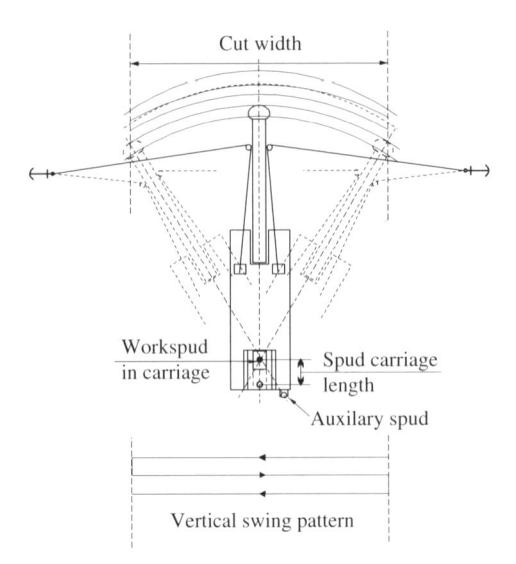

Figure 3.40. Swing pattern.

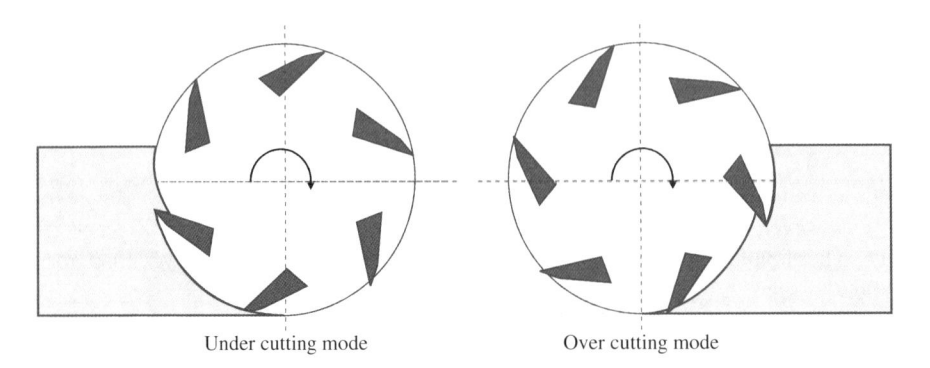

Figure 3.41. Cutting modes.

The swing from starboard to portside differs from portside to starboard due to the rotation direction of the cutter head; (over cutting) rotation in the direction of the swing movement or (under cutting) opposite to that (Figure 3.41).

Seagoing cutter suction dredgers have their own propulsion that is used only during mobilization. The propulsion is situated either on the cutter head side or on the spud poles side.

3.3.2.2 *Areas of application*

Cutter suction dredgers are largely used in the dredging of harbours and fairways as well as for land reclamation projects. In such cases the distance between the dredging and disposal or reclamation area is usually smaller than these distances covered by trailing suction hopper dredgers. The cutter suction dredger is very suitable when an accurate dredge profile is required.

The cutter suction dredger can tackle almost all types of soil, depending on the installed cutting power. Cutter suction dredgers are built in a wide range of types and sizes, the cutting

Figure 3.42. Seagoing CSD Marco Polo loading barges.

Figure 3.43. All Wassl Bay.

head power ranges between 20 kW for the smallest to around 7,000 kW for the largest. The dredging depth is usually limited; the biggest suction dredger can reach depths between 30 and 35 m. The draught of the pontoon usually determines the minimum dredging depth. Under offshore condition the workability, even for the self propelled CSD's is limited. In the past 2 offshore CSD have been build The All Wassl Bay and the Simon Stevin.

In the late seventies and early eighties of the previous century two offshore cutter suction dredgers have been built for applications offshore. The All Wassl Bay (Figure 3.43) build by Mishubitsi, Japan for Gulf Cobla Ltd in *Dubai*. The All Wassl Bay has dredged the approach channel to the harbour *Jebel Ali* in *Dubai*, Unit Arab Emirates. After 2 years working the dredger is sold and scrapped.

78 DREDGING IN COASTAL WATERS

Figure 3.44. Simon Stevin.

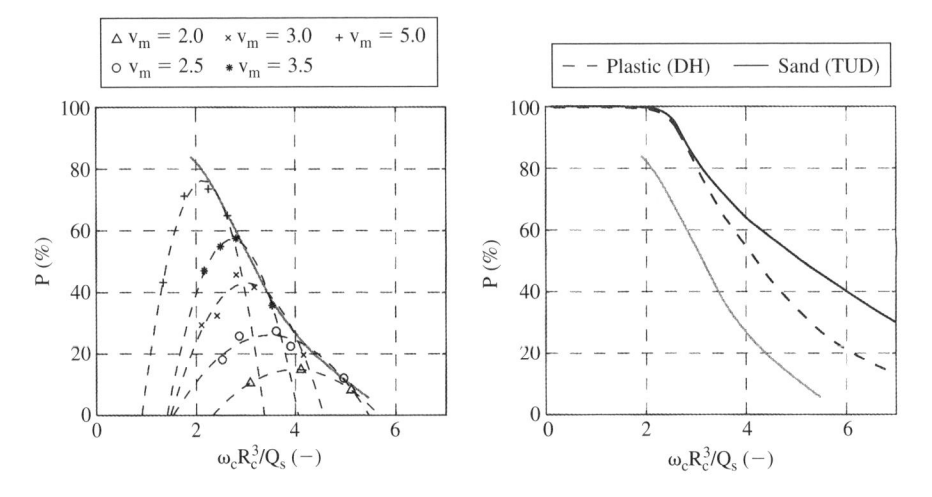

Figure 3.45. Production percentages versus inverse flow number for under cutting.

The Simon Stevin (Figure 3.44) build for Volker Stevin Dredging has even never worked. Both dredgers appeared too specialised to be economical.

3.3.2.3 *Environmental considerations*
The soil cut by the cutter head is mixed with water inside the cutter head, in order to make it suitable for hydraulic transport. Unfortunately not all the material cut, is picked up by the pump flow, but a part of the soil will leave the cutter head before it enters it the suction mouth. This is called spillage and is expresses as the ratio of the weight of the soil left to the total weight of the soil cut. It reduces the productivity of the dredger and can cause turbidity of the surrounding water.

Den Burger (2003), showed in his thesis that the production percentage (100% minus percentage spillage) depends on the flownumber and particle size. The flownumber is the ratio of two flows namely: the flow generated by the cutter head and the suction flow (Figure 3.45).

Figure 3.46. Plain suction dredger (PSD).

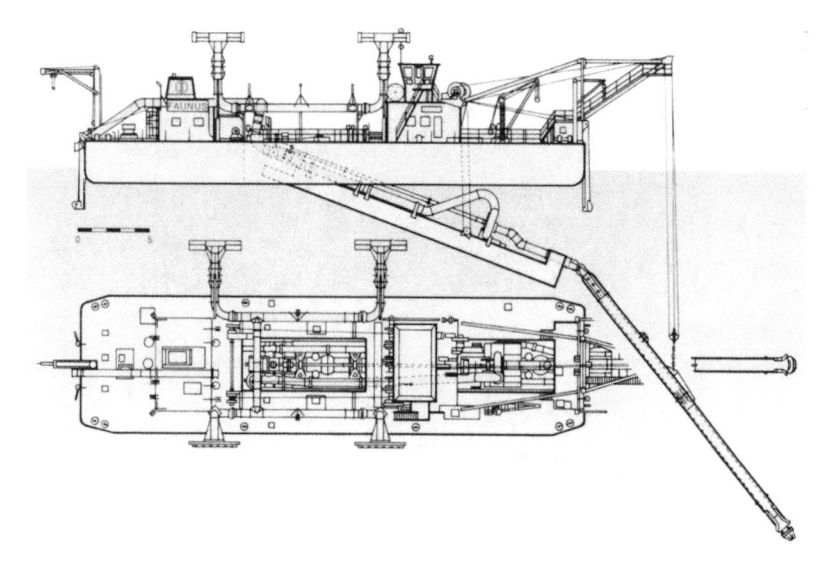

Figure 3.47. Plan view of a PSD.

For silt and soft clays turbidity due to this phenomenon can be minimised by reducing the flownumber to a sufficient low value. In some cases it might be necessary to change the dimensions of the cutterhead to achieve an acceptable spillage level (den Burger et al., 2004).

3.3.3 The plain suction dredger

3.3.3.1 *General considerations*
The characteristic of a plain suction dredger is that it is a stationary dredger, consisting of a pontoon anchored by on wires (mostly 6) and with at least one sand pump that is connected to a suction pipe. The discharge of the dredged material can take place via a pipeline or via a barge-loading installation. The suction tube is positioned in a well in the bow of the

80   DREDGING IN COASTAL WATERS

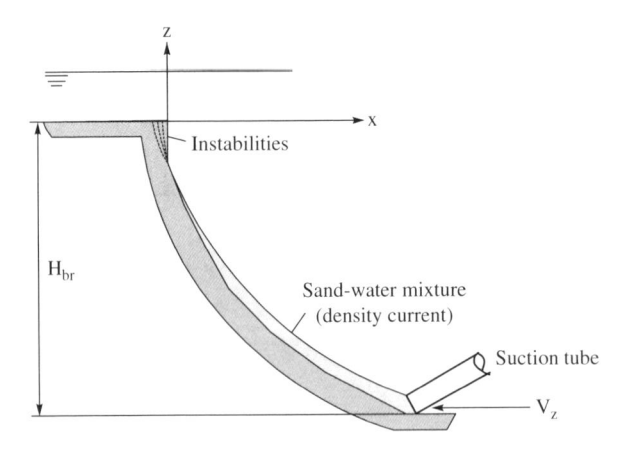

Figure 3.48. Breaching.

Figure 3.49. Jet system near the suction mouth of the PSD decima.

pontoon in which it is hinged. The other end of the suction pipe is suspended from a gantry or A-frame by the ladder hoist. The ladder hoist is connected to the ladder winch in order to suspend the suction pipe at the desired depth.

3.3.3.2 *General considerations*

The dredging method of the suction dredger is based on both the progressive collapsing of the breach or bank (Figure 3.48) and the loosening of the sand near the suction mouth by eddies created by the flow of water caused by the sand pump and jet pumps (Figure 3.49). The progressive collapse of the breach resulting from the dislodgement of particles of soil or masses of soil as a result of localised instabilities is termed "breaching". This process is essential for the production of a suction dredger and is entirely determined by the soil properties of the slope, the most important factors being its permeability and relative density of the soil.

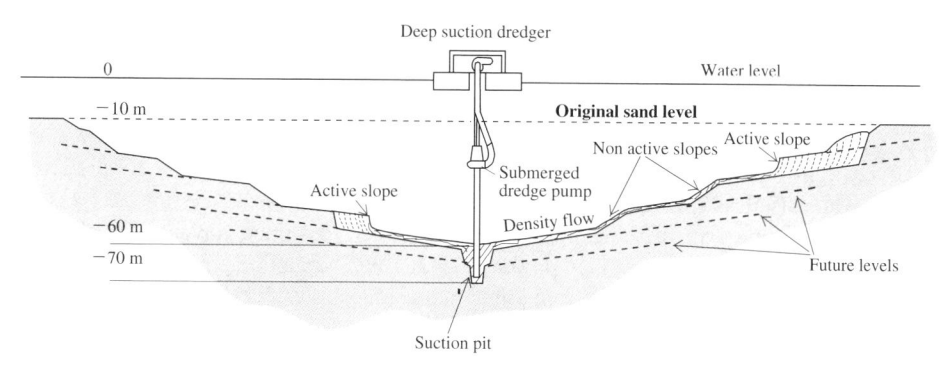

Figure 3.50. Deep suction dredging.

From investigation by De Koning (1970) it appeared that at the base of the breach a very gentle slope occurred, much smaller than the natural slope cause by the density current. It is even possible that sedimentation will take place at the base of the breach (Figure 3.50).

The flow of soil to the suction mouth depends strongly on the height of the breach. When the slopes reach the equilibrium values, the production reduces and the dredger is moved forward by a set of winches.

Almost all plain suction dredgers are equipped with water jets to improve the breaching process when necessary and/or to support the mixture forming process near the suction mouth in order to dredge the sand with the required density. It will be clear that improving of the breaching process is very local looking of the size of the pit in relation to the influence area of the nozzle. To control the mixture forming, a water jet is system is mounted around the suction mouth.

Although, in principle, the complete breaching process can be described with the existing erosion and sedimentation theories together with instability calculations of the slope, it will be clear that due to the variation of the soil, the existence of layers of cohesive material etc, the accuracy of the calculation is limited.

Borrow pits from plain suction dredgers are therefore very irregular in shape instead of having the shape of a cone. Due to the variation of the soil and the size of the pit some segments of the pit may breach quite different than others.

3.3.3.3 *Areas of application*

Plain suction dredgers are only used to extract non-cohesive material. Moreover these dredgers are less suitable for accurate work such as the making of specified profiles. Suction dredgers are very suitable for the extraction of sand, particular when the sand layers are thick (over 10 m).

If the dredger is equipped with an underwater pump, it is possible to dredge at depths exceeding 80 m. Depending on the pumping capacity and booster pumps, it is possible to transport material over considerable distances (30 to 50 km) via hydraulic pipelines.

Dredging with plain suction dredgers is more attractive when the production of the breach exceeds the production of the pumps of the dredger. When this is not the case, the breaching process has to be activated by powerful water jets. When the dredging depth allows it, a cutter suction dredger can be used to reach the higher required production level.

The latter has the advantage that more sand can be borrowed from the pit due to a more accurate dredging method.

Many suction dredgers are demountable; therefore they can also be used in excavation pits, which are not connected to navigable waterways. In general, suction dredgers are relatively light vessels and, although anchored on wires, are usually unsuitable for dredging in open waters (unless specially adapted).

Figure 3.51 shows an offshore plain suction dredgers designed for significant wave heights of 2.75 m and a total installed power of 7,425 kW. The suction pipe consists of three parts connect to each other by universal joints, which uncouples the ships movement from the suction mouth. The coupling with the floating pipeline is in the middle of the port side where the movements of the pontoon are minimal when working in waves.

The above system makes it possible to borrow industrial sand underneath overburdens, however the volumes of borrowed sand compared the volumes present are small (<30%).

In unprotected waters and long transport distances barge loading is more appropriate, because boosters at sea, although possible, are a source of troubles.

3.3.3.4 *Environmental considerations*

As said earlier plain suction dredgers are only used for dredging non cohesive soils. Density currents flowing from the breach cause local turbidity in the borrow pits. Barge loading causes overflow losses and therefore additional turbidity levels in water column.

Large abandoned suction pits can act as sinks for sediments originating from the surrounding areas and depending on the local flow and wave patterns.

Furthermore, short and long-term effects may occur on marine and coastal benthic communities of plants and animals (ecology). At sea such pits can also have undesirable consequences on adjacent coastlines.

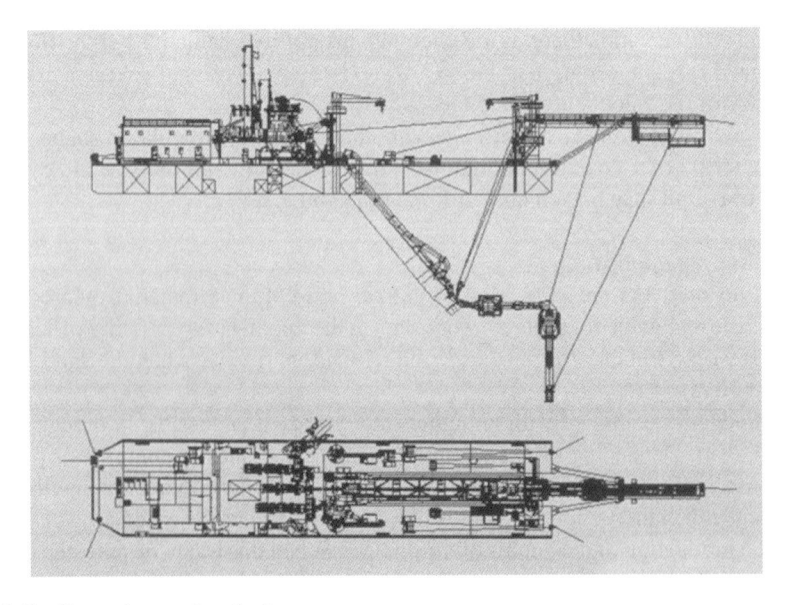

Figure 3.51. Sea-going suction dredger.

Due to the inaccuracy of the dredging process as well as that the plain suction dredger is only suitable for dredging non cohesive soils she is unsuitable for removal of contaminated soil.

3.3.4 The water injection dredger

In water injection dredging, a layer of suspended sediment in water with a substantially higher density to that of the surrounding water is produced (Figure 3.52). A pipe, with water nozzles arranged at small separations perpendicular to the direction of the motion, is brought over the sediment at a small distance from the bottom and a large volume of water is injected into the sediment under low pressure. Thus it is possible for a natural density current to occur, due to the difference in density between the mixture and the surrounding water. The sediment layer remobilized by the water injection may even spread against the direction of the tide and sometimes run off over sandbanks. The suspended sediment layer can flow into deeper areas of water and can be deposited again, or it may be guided into regions of high flow-rate and turbulence, in which further transport occurs. The achievable transport routes depend on the composition and density of the mobilised sediment and the local morphological and hydrological conditions.

When the sediment returns to its natural solids regime, which in tidal waters is characterised by strong dynamics i.e. deposition and resuspension of sediments and in part high natural concentration of suspended matter, control over the sediment is given up, and the sediment is subject to natural processes again. This means that exact knowledge of the local circumstances is necessary, so as not to gain unwanted sedimentation at other places.

The water injection process is generally described as a very effective technique and has shown itself to be extremely effective in tidal harbours, especially for fine-grained silty sediments. Most investigations show that the sediment material is not mixed into the upper water volume, and sediment transport is predominantly close to the bottom. The density currents produced are, as a rule, in the lowest 1–3 metres above the bottom. The solids

Figure 3.52. The water injection dredger.

Figure 3.53. Bucket ladder dredger.

concentration of the density current close to the bottom may be up to 100 g/l. Investigations in the Mississippi showed turbidity values close to the background value at a distance of about 240 m downstream of the dredger. The mixing of the suspended sediments in the vertical profile of the water depends substantially on the flow velocity in the local area and the composition of the sediment. Fine-grained particles are more easily mixed into the water column than sandy sediment material. In tidal waters, under some circumstances, mixing into the water volume may occur, but enhanced concentrations of suspended material have been observed only for short periods and distances. The mixed-in quantities of sediment are as a rule small compared with the natural sediment transport. Consequently quite different transport distances, from a few metres to a few kilometres, are achieved. The influencing factors include sediment properties, i.e. composition and density, morphology and flow in the wider area of the dredging operations. Silty bottoms are as a rule transported over relatively large distances, whilst for sandy bottoms the natural transport is substantially shorter. To achieve larger distances, the dredging operation may have to be repeated.

Since a real control of the sediment is absent practically all of the time, and particularly in tidal waters, it is possible to get a reverse transport of the dredged sedimentary material. However, in the investigations in Scotland, no significantly enhanced sedimentation was observed at any station.

The sediment composition of the bottom can be changed by WID operations. The sand component increases as a result of carry – off of fine-grains with the flow. Hence normal' dredging operations may need to be undertaken after repeated use of the water injection system.

3.3.5 The bucket ladder dredger

3.3.5.1 *General considerations*

A bucket dredger is a stationary mechanical dredger that is equipped with a continuous chain of buckets, which are carried through a structure, the ladder (Figure 3.55). This ladder is mounted in a U-shaped pontoon. The drive of the bucket chain is on the upper side. The bucket dredger is anchored on six anchors. During dredging, the dredger swings round the bow anchor by taking in or paying out the winches on board (Figure 3.54).

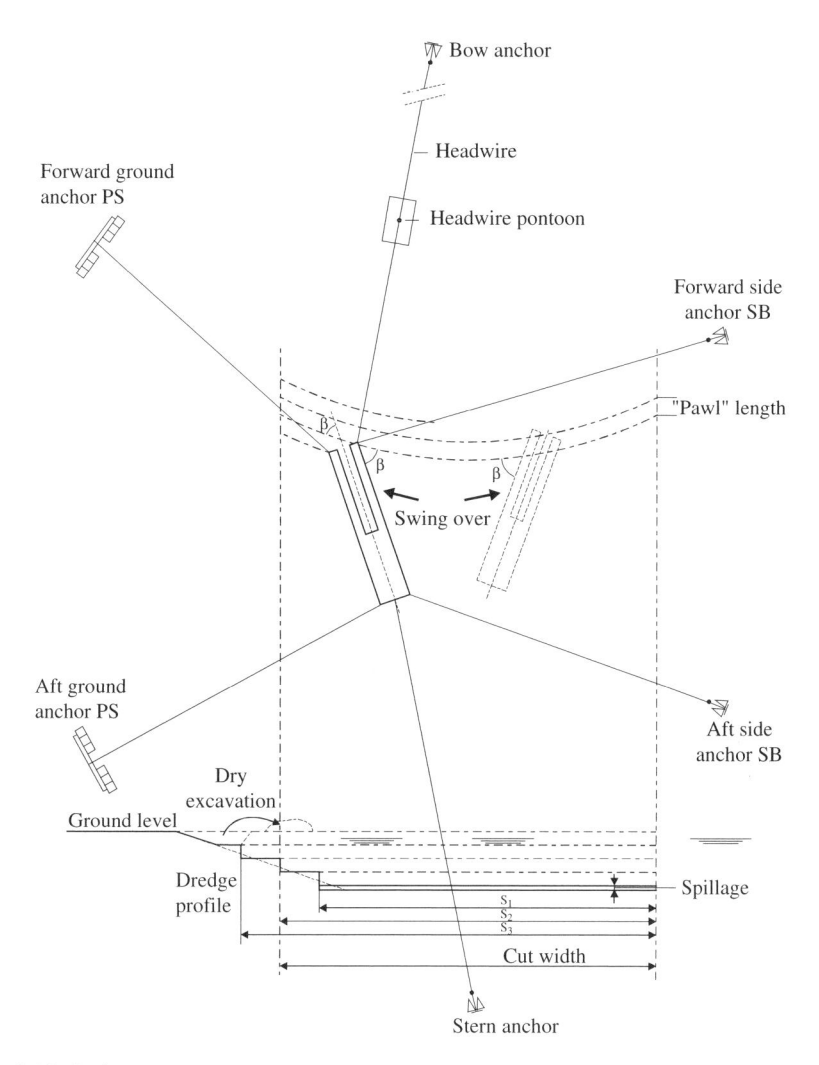

Figure 3.54. Swing pattern BLD.

The buckets (Figure 3.55 d), which are filled on the underside (e), are emptied on the upper side (b) by tipping their contents into a chute (h) along which the dredged material can slide into the barges (r) moored alongside. The chain is driven by the so called upper tumbler (d) at top of ladder frame, which is connected either via a belt to the diesel or directly to an electro motor or hydro-motor. The horizontal transport of the dredged material is mainly done by barges.

The capacity of a bucket dredger is expressed in terms of the content of the buckets. The capacity of a bucket can vary between 50 and 1,200 litres. Rock bucket dredgers often have a double set of buckets, the small rock buckets and the large mud buckets. This is in order to make better use of the power of the dredger and to widen the range of its use. The dredging depth varies between 10 m for the small ones and 35 m for the large dredgers.

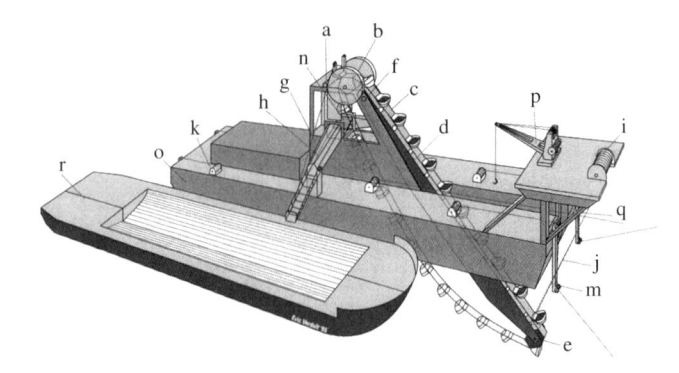

Figure 3.55. Main parts of a BLD.

Figure 3.56. Grab hopper dredger.

3.3.5.2 *Areas of application*
Bucket dredgers are only used in new (capital) or maintenance dredging projects when the initial depth of the area to be dredged is too shallow for trailing suction hopper dredgers and the distances involved are too long for hydraulic transport. For environmental projects, which require the dredging of "insitu densities", the bucket dredger is suitable peace of equipment.

When dredging for construction materials such as sand and gravel, or for minerals such as gold and tin ores, bucket dredgers are still frequently used.

Although this dredger is suitable to dredge all kinds (contaminated) sediments she can't be applied under offshore conditions. Furthermore, this dredger is expensive in maintenance cost and therefore replaced by the Backhoe Dredger in many cases.

3.3.5.3 *Environmental considerations*
Fluid soils can be eroded out off the buckets on its way from the seabed to the waterlevel. Besides cohesive materials can stick to the bucket walls and can fall out off the bucket in the water on its way back to the seabed. Although the bucket ladder dredger excavates the material with almost insitu density, the buckets carry always a certain percentage of water. This water is dumped into the barge too. Some barges do have possibilities to remove

Figure 3.57. 200 m^3 grab.

Figure 3.58. 30 m^3 grab dredging within a silt screen.

the water from the sediment via an overflow. This will introduce some turbidity in the surrounding of the dredger.

So, to avoid these problems special arrangements are required to avoid diluting of the water.

3.3.6 The grab dredger

3.3.6.1 *General considerations*

The grab dredger is the most common used dredger in the world, especially in North America and the Far East. It is a rather simple and easy to understand stationary dredger, with and without propulsion. In the latter the ship has a hold (hopper) in which it can store the dredge material. Otherwise barges transport the material. The capacity of a grab dredger is expressed in the volume of the grab. Grab sizes varies between less than 1 m^3 up to 200 m^3 (Figure 3.57).

The opening of the grab is controlled by the closing and hoisting wire or by hydraulic cylinders (Figure 3.59).

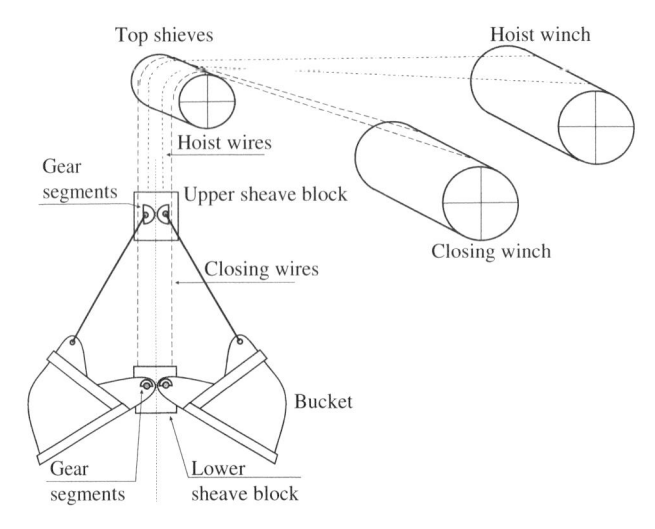

Figure 3.59. Hoisting and closing system of a cable grab dredge.

3.3.6.2 *Areas of application*
The large grab dredgers are used for bulk dredging. While the smaller ones are mostly used for special jobs, such as:

- Difficult accessible places in harbours
- Small quantities with strongly varying depth
- Along quay walls where the soil is spoiled by wires and debris
- Borrowing sand and gravel in deep pits
- Deep offshore dredging

The production of a grab depends strongly on the soil. Suitable materials are soft clay, sand and gravel. Though, boulder clay is dredged as well by this type of dredger. In soft soils light big grabs are used while in more cohesive soils heavy small grabs are favourable.

Grab cranes mounted on vessels are suitable for dredging under offshore conditions, however the accuracy of the positioning of the grab or clamshell requires much attention. From that point of view hydraulic cranes (Figure 3.60) are favourable above cable cranes, however the length of their boom limits the dredging depth. For deep offshore dredging work positioning the clamshell by thrusters gives a solution for the required accuracy (Figure 3.61).

3.3.6.3 *Environmental considerations*
Standard grabs are open at the top, therefore dredge material can be eroded the flow generated when the bucket is hoisted, causing spillage and turbidity. In soft soils bucket filling degrees can be more than 100% causing spillage on its way to the sealevel and to the barge. Furthermore most of these buckets cause an irregular seabed because the buckets don't have horizontal closing curves.

Figure 3.60. Hydraulic crane.

Figure 3.61. Controllable grab.

Figure 3.62. Working vessel for deep offshore dredging.

3.3.7 The backhoe dredger

3.3.7.1 *General considerations*

A backhoe is a stationary tool, anchored by three spuds: two fixed spuds at the front and a moveable spud at the back of the pontoon (Figure 3.63 and Figure 3.66). At the front of the pontoon in general a standard crane is mounted on a rotary table, of a well-known brand (Demag, Liebherr, O&K Poclain, etc.). Bucket sizes from several cubic metres to 20 m^3 exist.

Hydraulic cranes are available in two models, the backhoe and the front shovel. The first is used most.

The difference between those two is the working method. The backhoe pulls the bucket to the dredger, while the front shovel pushes. The last method is only used when the water depth is insufficient for the pontoon.

These stationary dredgers are anchored by three spud poles; two fixes to the front side of the pontoon and one movable at the aft side. This means that the dredging depth is limited

Figure 3.63. Backhoe dredger.

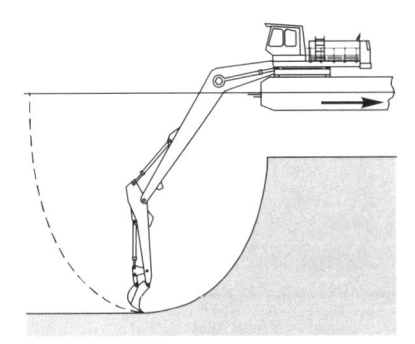

Figure 3.64. Backhoe mode. 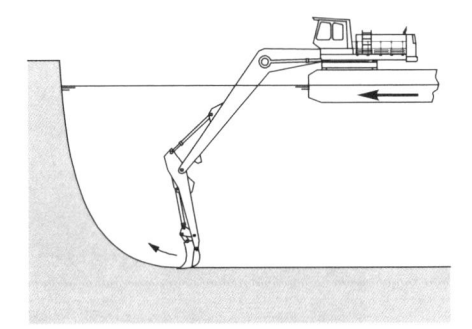　　　　　　Figure 3.65. Front shovel mode.

to about 20 m (maximum 25 m). During dredging the pontoon is lifted somewhat out of the water by wires running over the spud poles. In this way, a part of the weight of the dredger is transferred via the spuds to the bottom, resulting a sufficient anchoring to deliver the required reaction for the digging forces. Besides that the dredger is in this case less sensible for waves. At the front of the pontoon is normally a standard cranes mounted. Here pontoon deck is lower to increase the dredging depth. Bucket sizes vary from a few m^3 to 20 m^3.

This type of dredger is used frequently in the Scandinavian Coastal waters.

3.3.7.2 *Areas of application*
Backhoes are used in all soil types but particular in firm clay, soft rock, blasted rock and when large stones can be expected, like on slopes of waterside protections or in rocky watersides. The length of the stick and the boom determines the dredging depth (Figure 3.67).

Some backhoes have more than one bucket to be able to dredge well at several depths, a large bucket for soft soils and a smaller bucket for deep water or heavy soils. The lack of anchorage cables limits the hindrance for other ships and there is also no delay for anchorage. Hydraulic backhoes are especially suitable for accurate dredge work, due to

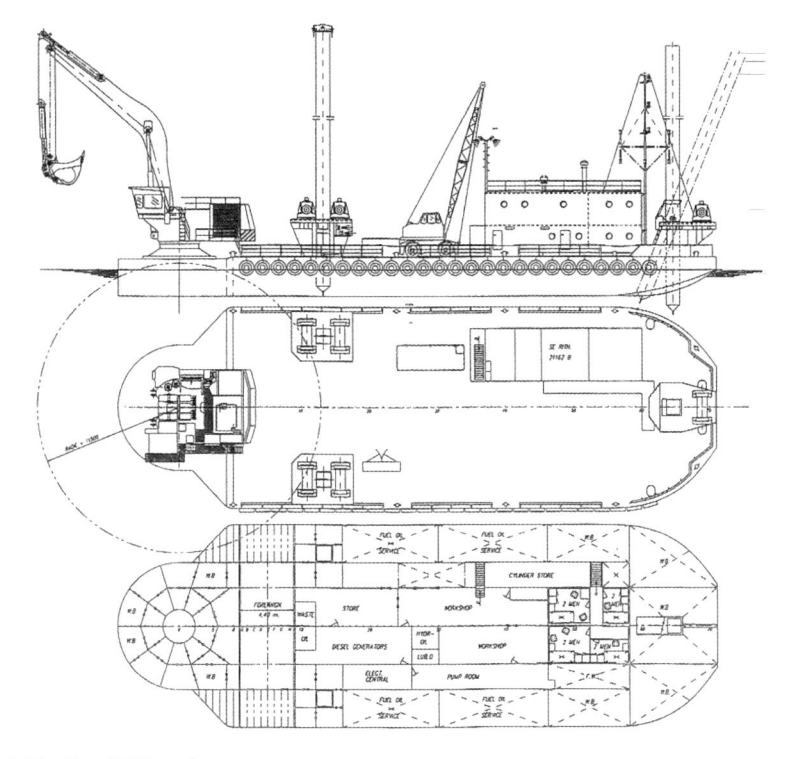

Figure 3.66. Plan BHD rocky.

the construction of the stick and beam. For environmental dredging special buckets are developed (Figure 3.82) to reduce induced turbidity when the bucket is lifted.

In general this dredge tool cannot be used on rough open water, due to the limited pontoon width.

3.4 ENVIRONMENTAL ASPECTS WHEN DREDGING

Distinction should be made between dredging contaminated sediments and non-contaminated sediments. The purpose of dredging contaminates sediments is either to maintain a certain navigational depth and/or to remove the contaminants out of the environment. In both cases distribution of the contaminants during dredging has to be avoided. Dredging of non-contaminated sediments can have environmental effects too, f.i. due to the removal of the sediments, inducing high turbidity levels, etc.

The vulnerability of a coastal sea to contamination is determined by its physical characterization. Seas are large and the contaminants deposed into the sea by the rivers are very much diluted. What happens after the deposition depends strongly on the water depth of the coastal sea and the flushing time of the sea with the ocean. The North Sea has a flushing time of several years, but it is a shallow sea with dynamic sediment behaviour.

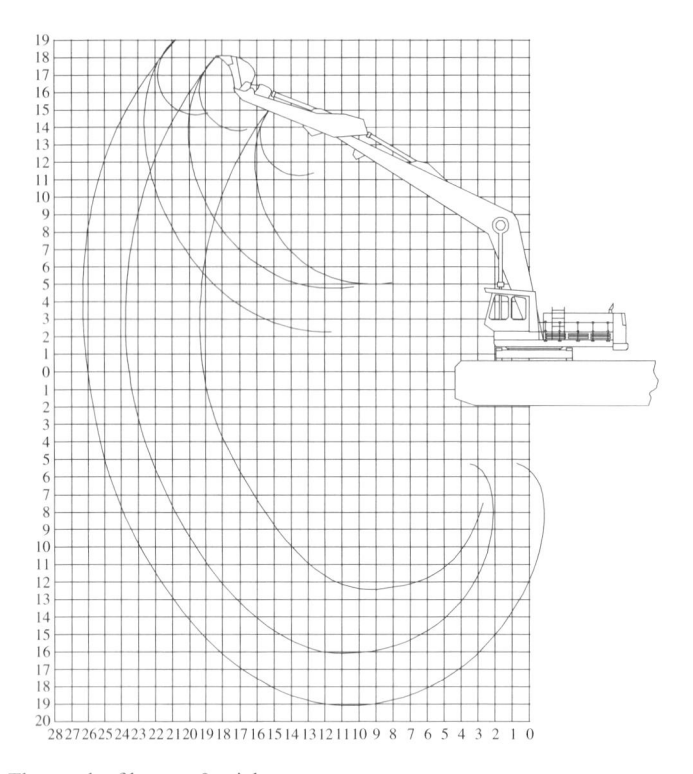

Figure 3.67. The reach of booms & sticks.

Re-suspension of sediments occurs on large scale during strong tidal currents and storms, which settles during neap tides or quiet weather periods.

Examples of the effect are shown in (Figure 3.14 and Zhang, 2003).

Figure 3.68 shows the distribution of lead in surface sediments from the North Sea. The lead concentrations are determined in the fraction smaller than 20 micron. High concentrations were found in the German Bright and in the central North Sea (upper left). The first is due to deposition of particle bound lead discharged by Belgian, Dutch and German rivers and transported by the prevailing currents to this location. The high concentration in the central part may be from atmospheric deposition (Aquatic Pollution, 1990). The silt content (smaller than 63 micron) in the surface sediments of the North Sea is mainly lower than 5%.

Figure 3.69 shows PCB 138 concentrations in the North Sea. The highest input into the North Sea is in front of the Dutch coast. Currents transport the sediments to the North, where a significant part settles in the Wadden Sea (Aquatic Pollution, 1990). Surface sediments in the Wadden Sea have a much higher silt content than those at the north Sea.

Estuaries are, apart from rivers and seas, contaminated by local sources and vary therefore strongly from case to case. The amount of dredging depends on the sediment balance and by the human activities necessary to keep the shipping lanes open. Highly polluted areas are found in the Northern Delta basin in The Netherlands.

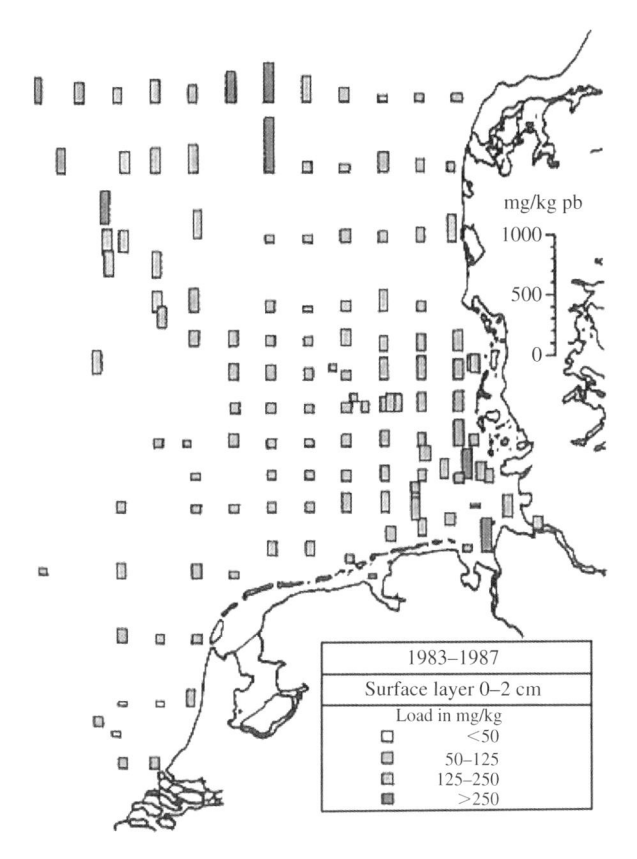

Figure 3.68. Distribution of lead in the North Sea.

For the removal of contaminated sediments special equipment has been designed or existing equipment adapted. A well-designed environmental dredger fulfils the following requirements:

- It can dredge a high concentration of sediments, preferable close to the insitu densities.
- It can dredge very accurate in the horizontal as well as in the vertical plane (± 10 cm).
- It induce a no or a few additional turbidity to the surrounding water.

In principal these requirements are applicable for non-environmental dredgers too, but due to the high cost of storage or cleaning of the contaminated sediments, the requirements are of higher importance for environmental dredgers.

All the dredger types mention in Section 3.3 "Dredging Equipment" are use for the removal of contaminated sediments, sometimes special-designed or adapted or just plain. Which type of dredger is the best applicable for the removal of contaminated sediments depends on:

- The volumes to be dredged.
  Very large volumes such as encountered in the Port of Rotterdam are mostly removed by trailing suction hopper dredgers.

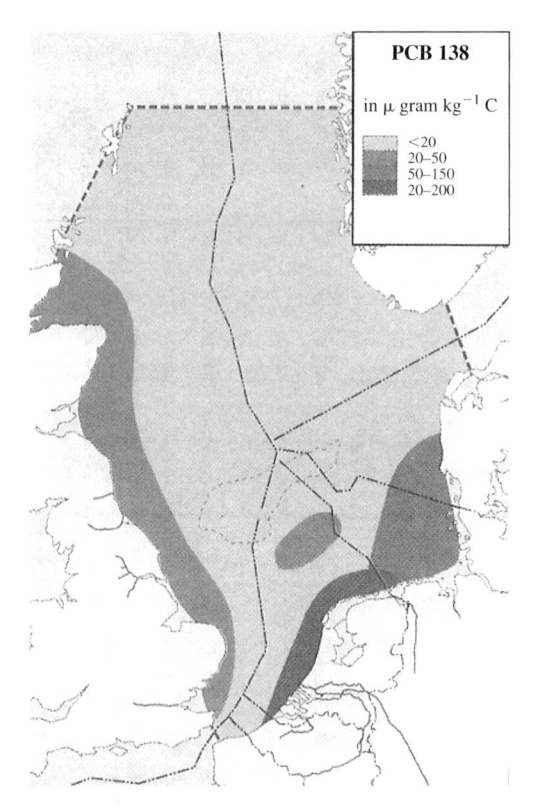

Figure 3.69. Distribution of PCB in the North Sea.

Sophisticated environmental dredgers are cost effective when large volumes have to be removed. Small volumes are mostly dredges with grab or backhoe type dredgers. For the cleaning of ponds or small canals or moats simple dredgers or even shovel type dredgers are used.
- Access to the site.
  If the dredging site is not accessible via water, dismountable dredgers are required, which mostly reduced outputs.
- The dredgeability of the sediment.
  Is the silt cohesive or behave it as a fluid. In other words is a cutting device required?
- The thickness of the layers to be removed.
  Relative thin layers require a special cutting and/or suction head to avoid diluting of the sediment with the surrounding water.
- The amount of debris in the sediments (Figure 3.70).
  Rotating cutting devices can't cope with wires and tires and much other debris.

3.4.1 The trailing suction hopper dredger

This type of dredger is mainly used in its plain state. The minimum requirements are a pump ashore equipment to pump the silt to (confined) disposal areas and a degassing installation.

Figure 3.70. A barge full of debris.

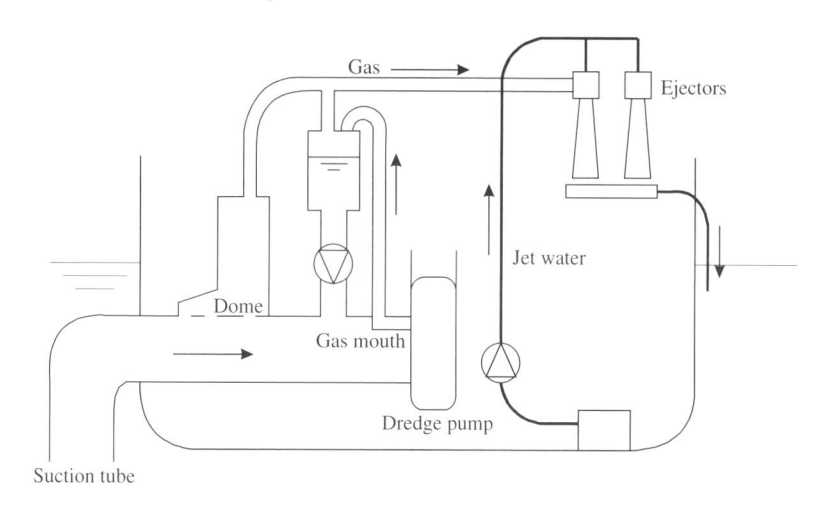

Figure 3.71. Scheme of a degassing installation.

The last one is necessary to fulfill the condition of high concentrations and a regular suction process. Silts contain frequently gas from organic materials. When dredging these silts, gas bells will grow due the reduction of the pressure in the pipeline on their way to the dredge pump. When these bells are not removed, their effect is equal to that of cavitation bells and the performance of the pump reduces rapidly with the bell volume. A degassing installation (Figure 3.71) removes these gas bells out of the flow, which results in a normal pump performance and so in higher solid concentrations dredged. The gas is assembled in a dome

Figure 3.72. The auger dredger HAM 211.

or accumulator from which the bottom is connected to the suction pipe by special shaped holes. A secondary removal of the gas is in the impeller eye. The gas is removed by ejectors which are driven by jet water. The gas is mixed with a large amount of jet water and can be released safety in the air however, discharging below the water level is preferable.

Overflowing of contaminated sediments should be avoided at all times so when the load in the hopper reaches the level of overflow the dredging is stopped. Deposition of contaminated sediments at sea is not allowed, therefore the loads are pumped to areas where, if necessary, the sand content can be separated from the mass volume and the contaminated silt either immobilized or stored in a confined disposal area.

As said the TSHD for the removal of contaminated silt is used mainly in busy ports and large quantities and sufficient thick layers. The vertical positioning of the draghead is for bulk dredging about 0.4 m. The horizontal positioning is much less therefore, when all the material has to be removed, the last part is done by another, more accurate working type of dredger.

3.4.2 Cutter suction dredger

Many environmental dredgers are based on this type of dredger, because it can fulfill the best the accuracy criteria of the excavating tool, while the dredged material can directly be transported by pipeline to a disposal area when applicable. However, as shown above, the cutter head is not the ideal cutting device to minimize induced turbidity due to dredging. Therefore the cutter head is replace by other excavating tools such as:

A. An auger
B. The bottom disc cutter head
C. The scraper head
D. The sweep head
E. The slice head

Ad A. The auger dredger (Figure 3.72).
When dredging, the auger shaft is parallel with the centreline of the dredger and is moved from starboard to ports and vice versa. The auger (Figure 3.73) transports the material to its centre where the sediments are picked up by the pump flow. A movable cap that closes

Figure 3.73. The auger.

Figure 3.74. The bottom disc cutter.

the top and backside of the auger prevents induced turbidity of the surrounding water by the auger; the minimum layer thickness which can be dredged is 2 cm!

When changing from swing direction the auger is turned 180 degrees.

The particular dredger shown in Figure 3.72 has a possibility to control the position of dredger to the movable spud in such away that the dredger is able to make parallel cuts instead of concentric cuts. The spillage induced by an auger is very low.

Ad B. The bottom disc cutter (Figure 3.74).
The bottom disc cutter is a cutting device having low spillage capabilities and is particular suitable for cutting soft clay and silts. A vertical movable cover closes the non-cutting side of the head and the opening on the cutting side above the sediment to reduced spillage and induced turbidity. However, when changing the swing direction the cover is switch to the other side. The thickness of the layer to dredge can be varied from zero to eighty cm. should roughly be equal to the height of the disc.

98 DREDGING IN COASTAL WATERS

Cross section pushing

Cross section pulling

Top view

1. Visor
2. Scraper (when pushing)
3. Scraper (when pulling)
4. Levelling plate
5. Hinge joint

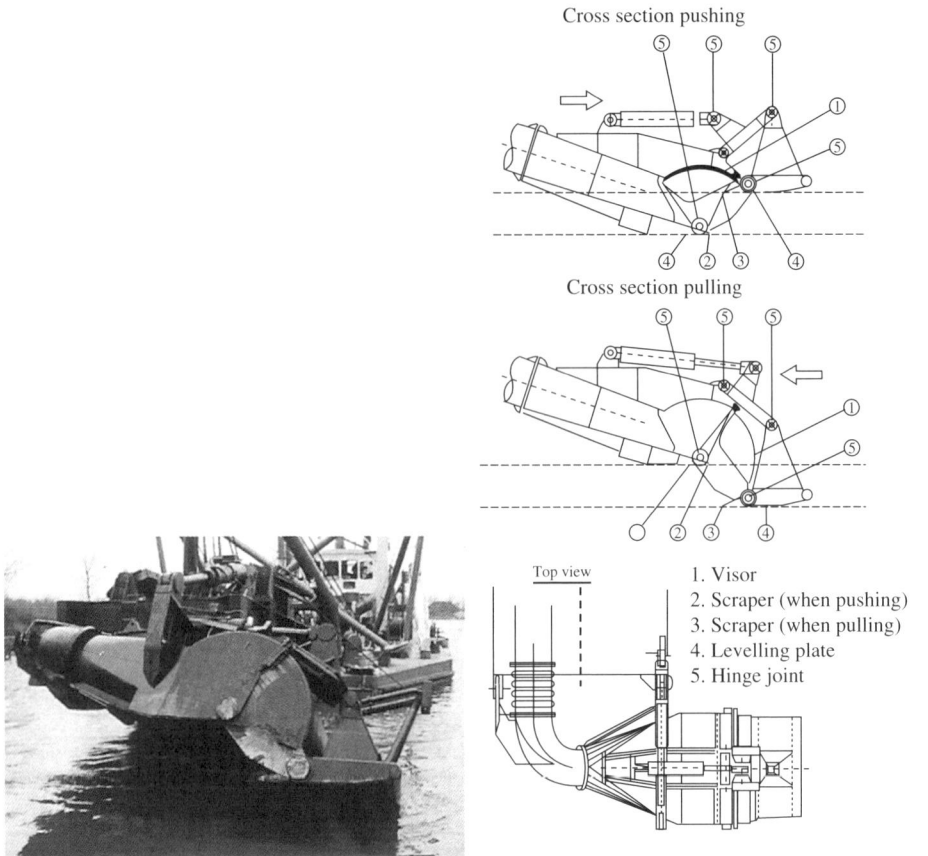

Figure 3.75. The sweep head.

Figure 3.76. Cross section and top view of the sweep head.

Ad C. The sweep head (Figure 3.75).
The sweep head looks like a draghead of a trailing suction hopper dredgers, it is dragged to one side and pushed to the other side (Figure 3.76). When dragged the visor is turned down to close the backside of the head, when pushing the visor up and the leveling plate guides the sediment layer into the head and reduces the entering of the water flow. In both cases the sediment is pushed into the draghead and hydraulically transported.

The draghead is direct constructed to the suction pipe which enables an accurate horizontal and vertical positioning. Adjustable water injection in the sweep head enables control over the density and viscosity of the mixture.

Ad D. The scraper head (Figure 3.78).
The scraper head is a trapezium shaped excavating device, covered by a fixed cap and can be used in both swing directions. The cap is controlled by a hydraulic cylinder (Figure 3.77).

Ad E. The slice head (Figure 3.79).
The slice head is a horizontal funnel shaped head partly open at the underside and connected to the suction pipe of the dredger. The slice head is pulled by the side winch wire of the

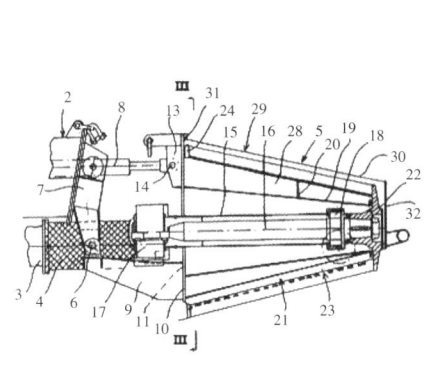

Figure 3.77. Cross section over the scraper. Figure 3.78. The scraper head.

Figure 3.79. The slice head.

dredger trough the silt layer. Because of the funnel shaped head the silt is driven up and closes the aft side of the head over the full height to avoid the suction of surrounding water. When necessary mixers can be placed into the head to reduce the viscosity.

The slice head can be used only in the pushing direction and don't have any cutting device.

3.4.3 Bucket ladder dredger

The bucket ladder dredger dredges the material with densities close to the situ-density. However, transporting the soil from the seabed to the water level the soil in the bucket is in contact with the surrounding water and causes an increase in turbidity.

Underwater tunnels are place over the bucket ladder (Figure 3.80), to avoid possible spill and to reduce induced turbidity of the water. Special valves in all buckets are mounted to

Figure 3.80. An environmental BHD.

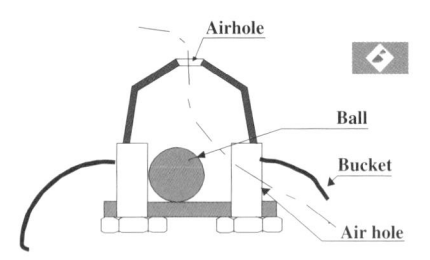

Figure 3.81. The air valve in the bucket.

enable air entering (Figure 3.81) to assist the emptying of the buckets or letting air leaving the buckets on its way down to the seabed. To avoid stick material to the buckets the buckets are leaned by water jets just after passing the upper tumbler. Depending of the bucket ladder angle more or less water is transported with the sediment into the barge.

3.4.4 Backhoe dredger

As said earlier, hydraulic backhoes are especially suitable for accurate dredge work, due to the construction of the stick and beam. For environmental dredging special buckets are develop to reduce induced turbidity when the buckets are lifted in the water (Figure 3.82). For dredging thin sediment layers special so called visor buckets are developed, which can closes the bucket by a movable visor.

The disadvantage these visor buckets is that they can only close well when there is no debris in the soil to be removed.

A rubber cover is in that case less sensible, but closes badly (Figure 3.83).

3.4.5 Grab dredger

When adapted, grab dredgers can be used for environmental dredging too. In that case the grabs should be closed at the top side, to avoid induced turbidity of the water when hoisted. Furthermore the required vertical accuracy can be obtained by using, so called, horizontal

Figure 3.82. Well covered bucket.

Figure 3.83. Rubber cover.

closing grabs. (Figure 3.84) This means that the closing curve of the grab, the line made by the cutting edge of the buckets parts, is (almost) a horizontal line. This enables dredging of relatively thin sediment layers with a sufficient filling degree. Horizontal closing grabs are available for hydraulic closing (Figure 3.66) as well as for cable closing (Figure 3.85 and Figure 3.86).

102 DREDGING IN COASTAL WATERS

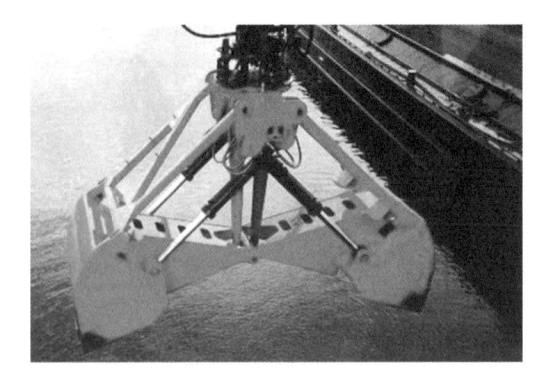

Figure 3.84. Horizontal closing cable grab.

Figure 3.85. Horizontal closing cable grab (top view).

Figure 3.86. Horizontal closing cable grab (top view).

Accurate positioning and monitoring system is required to realise an accurate grab dredging operation. Particular the horizontal positioning is better with hydraulic cranes than with cable cranes, in particular when dredging is required under current conditions.

For the bucket ladder, backhoe and grab dredger the transport of the dredged sediment, including some added water, is realised by barges.

Figure 3.87. Horizontal positioning.

As discussed above, during dredging the fine (contaminated) materials can be brought into the water column and causing turbidity. By using silt screen (Figure 3.58) dispersion of the turbidity can be reduced. However, tests in the Netherlands have shown that those screens around a dredging job are vulnerable, difficult in operation and in some case slightly effective.

Disadvantages of silt screens are:

- In tidal areas the screen depth has to be adapted regularly to the actual water level and the effect is reduced by the tidal currents.
- In current or flows induced by ship silt screens can be kept hardly vertical.
- For the effectiveness of the silt screen it is important that other turbidity sources (f.i. ships) are minimised too: the repositioning of the screen is time consuming and has influence on the price level of the project.
- Silt screens are vulnerable and require costly repairs.

Silts screen are useful to avoid dispersion of contaminated sediments when only a small volume water is subject to a high turbidity level, f.i. in quiet waters.

The cost of environmental dredging is much higher than for ordinary nautical dredging. The cost depends on the level of contamination, the percentage of sand in the sediment, possibility of cleaning, and deposition of the rest volume.

BIBLIOGRAPHY

Association of Dutch Dredging Contractors, Aquatic Pollution and Dredging in the European Community, DELWEL Publishers, The Hague, 1990, ISBN 90-6155-430-6

Marco den Burger, W.J. Vlasblom and A.M. Talmon, "*Influence of Operational Parametres on Dredge Cutterhead Spillage*" CEDA Dredging Days, 1999 November 18–19, Amsterdam

Marco den Burger, Mixture Forming Processes in Dredge Cutter Heads, PhD Thesis Delft University of Technology. 2003

J.B. Herbich, Coastal & Deep Ocean Dredging, Gulf Publishing Company, Houston Texas U.S.A, 1975, ISBN 0-87201-194-1

IADC & CEDA, Environmental Aspects of Dredging Guides, International Association of Dredging Companies (IADC), The Hague

J. de Koning, Neue Erkenntnisse beim Gewinnen und Transport von Sand im Spülproject Venserpolder, V.D.I. Tagung "Bauwen im Ausland", Hamburg 21–25, April 1970 (in German)

V. Van Lancker, Sediment and morphodynamics of the Belgian western coastal zone. Geologica Belgica, scientific seminars. Brussels, ULB, 6/04/2000 ...

Vera R.M. Van Lancker, Steve D. Honeybun and Geert P.M. Moerkerke, Sediment transport pathways in the Broers Bank – Westdiep coastal system. Preliminary results, Marine Sandwave Dynamics, International Workshop, March 23–24 2000 (eds. A. Trentesaux & T. Garlan), pp 205–212. University of Lille 1, France. Proceedings, 240 p (poster presentation)

Vaclav Matousek, Flow Mechanism of Sand – Water Mixtures in Pipelines, PhD Thesis Delft University of Technology. 1997, ISBN 90-407-1602-1

S.A. Miedema, The cutting process in Sand and Clay, Delft University of Technology, Section Dredging Engineering, Down loadable from WWW.Dredgingengineering.com

NATIONAL RESEARCH COUNSEL, Contaminated Sediments in Ports and Harbours, National Academic Press, Washington, D.C., 1997

OSPAR Commission 2000, Quality Status Report 2000, Region II – Greater North Sea

G.W. Plant, C.S. Covil and R.A. Hughes, Site Preparation for the New Hong Kong International Airport, Thomas Telford, 1998, of Sand – Water Mixtures in Pipelines, PhD

Cees van Rhee, On the Sedimentation in a Trailing Suction Hopper Dredger, PhD Thesis Delft University of Technology. 2002

Vera R.M. Van Lancker, Steve D. Honeybun and Geert P.M. Moerkerke, Sediment transport pathways in the Broers Bank – Westdiep coastal system

Han Winterwerp, On the Dynamics of High Concentrated Mud-Suspensions, PhD Thesis Delft University of Technology. 1998

CEDA Dredging Days, 1999 November 18–19, Amsterdam

W.J. Vlasblom, Deep Dredging Techniques, Proceedings Johannes de Rijke Memorial Symposium, page 394–402, Osaka, Japan 2000

W.J. Vlasblom, Production Control and Quality Control Aspects During Executions of the Reclamation Works at Chek Lap Kok. Terra et Aqua, March 1999, issue 74

W.J. Vlasblom, Investigations to the spillage of the horizontal suction process, Proceedings of the First International Dredging Congress and Exhibition, Shanghai China, 2003

W.J. Vlasblom, Designing Dredging Equipment, Lecture notes, Delft University of Technology, Section Dredging Engineering, Down loadable from WWW.Dredgingengineering.com.

W.J. Vlasblom, The Breaching Process, Delft University of Technology, Section Dredging Engineering, Down loadable from WWW. Dredgingengineering.com

W.J. Vlasblom, Dredge pumps, Delft University of Technology, Section Dredging Engineering, Down loadable from WWW. Dredgingengineering.com

W.J. Vlasblom, Cutting of Rock, Delft University of Technology, Section Dredging Engineering, Down loadable from WWW. Dredgingengineering.com

Qinghe Zhang, Lin Quanhong, Gu Ming and Li Jinjun, *Remote Sensing Image Analysis on Circulation Induced by Breakwaters in the Huanghua Port*, Proceedings of the International Conference on estuaries and Coasts, November 9–11, 2003 Hangzhou, China

WEB SITES

- http://sandpit.wldelft.nl/mainpage/mainpage.htm
- www.sandandgravel.com
- http://www.mumm.ac.be
- www.nitg.tno.nl

Chapter 4

Dredging in UK Coastal Waters

Neville Burt
HR Wallingford, Wallingford, UK

4.1 HISTORY

The history of dredging in the UK was reviewed by Dr John Riddell in a paper presented at the World Dredging Congress in Amsterdam (Riddell, 1995). The authors are grateful for his permission to quote extensively from that review in this book.

To an island trading nation such as the UK, sea transport has always been of vital importance. Despite the immense growth in air travel and the completion of the Channel Tunnel, the UK's ports and harbours were handling around 500 million tonnes of raw materials, foodstuffs and manufactured goods in the mid 1990's. This trade passes through some 150 separate ports of widely ranging size and ownership. Some have declined and some have increased. The overall trend in 2000 was still an increase in trade. Many of the ports are of considerable age and when first developed were situated in the sheltered waters of estuaries and coastal inlets. These sites had adequate natural depths of water for the vessels of their time, but as ship sizes increased and trade expanded rapidly with the Industrial Revolution the need for deeper water also increased. Early efforts to achieve greater navigable depths in estuaries and rivers concentrated on raising water levels by the construction of dams and locks. The alternative of lowering the bed level by dredging was soon found to be a more practical solution.

At first man and horse-powered dredging machines made some impact but it was the arrival of the steam engine that allowed real progress to be made. John Grimshaw's application of a 4 hp Boulton & Watt beam engine to a bag and spoon dredger at the Port of Sunderland in 1798 is believed to be the first recorded use of steam power for dredging. Five years later the first steam-driven bucket ladder dredger started work in Deptford Dockyard on the River Thames (Gower, 1968). Reputed to have achieved outputs of up to 120 tonnes of sand per hour from a depth of 5 m, it was the predecessor of a vast fleet of steam-powered bucket ladder dredgers which by the middle of the nineteenth century were to be found at work in virtually every harbour in the country.

The steam dredger was employed to create deeper water and for a variety of other harbour construction tasks. It was used to dig canals, excavate new docks and tidal basins and form the foundations for quay walls, breakwaters and bridges. Such capital dredging projects, large and costly as they were, went mostly unseen. Being mainly underwater, the only real evidence of the dredging was the steady increase in the size of ships using the harbour where the activity took place. To take the example of Glasgow, a port created entirely by dredging, the draft of ships able to navigate to and from the city increased from 3.5 m in 1824 to 6.7 m in 1864, to 9.0 m by 1914 and to 12 m by the time the great passenger liner "Queen Mary" passed downriver in 1936 (Riddell, 1982). All of this increase in depth, together with the associated widening and straightening of the 29 km long channel, was undertaken by the steam dredger. The benefits of the dredger to Glasgow and to those who lived and worked in the surrounding area were incalculable. Similar benefits resulted elsewhere around the UK coast as the dredger created the water depths which allowed shipping and trade to flourish.

Deepening of a waterway can upset Nature. To maintain the increasing depths and breadths in the many ports, still more dredging was required to remove the deposits of sand and silt. Thus arose the need for maintenance dredging. This is an activity which many harbour authorities soon realised was to form a significant proportion of their operating costs. Unlike the major capital dredging projects, maintenance dredging required to be undertaken on a frequent and regular basis. In the larger UK ports such as Liverpool, London, Hull and Glasgow, maintenance dredging developed into a continuous operation with fleets of dredgers working all the year round. Some ports were more fortunate and required dredging on a less frequent basis, but very few in the UK ever escaped the need for some expenditure on this activity. Even the country's smallest cargo and fishing harbours and ferry ports experienced a degree of siltation. This sand and mud was removed by dredging. Not only were the depths in the harbours maintained by the dredger – the benefits gained by the local communities from the increased depths also continued.

The combination of capital dredging to support new developments and maintenance dredging to keep the waterways deep resulted in a spectacular growth of UK dredging activity throughout the nineteenth century. For reasons of convenience, economy and independence most ports developed in-house dredging fleets. Harbour authorities such as the Clyde Navigation Trust, the Mersey Docks and Harbour Board, the Tyne Improvement Commissioners and of course the railway companies thus became owners of large numbers of dredgers. Most of these machines were of the traditional bucket ladder type, but towards the end of the nineteenth century the sand pump dredger made its appearance. Moving the material in suspension by powerful steam driven pumps, the suction dredger was particularly useful for creating new land for port and shipyard development through hydraulic fill. This type of dredger was soon making a major contribution to maintenance dredging in the approach channels to ports such as London, Liverpool and Preston. Even in the smaller ports it was common to find a dredger owned by the harbour authority. Constantly part of the harbour scene, its activity was observed and recognised not only by the fishermen and merchants who used the harbour but by the summer holidaymakers and visitors.

The need for dredging equipment by the UK's many ports resulted in the establishment of a strong dredger building industry. This specialised part of shipbuilding was further reinforced by the steady demand for dredgers from the countries of the British Empire. India, Australia and Canada all required dredgers of one type or another, as did the many smaller colonies. The UK's dredger building industry, as in other countries, concentrated in one small area. By the start of the twentieth century the town of Renfrew, near Glasgow, became synonymous with dredge construction. The neighbouring shipyards of William Simons and Frederick Lobnitz were always in keen competition, but for nearly a century created highly skilled employment and many innovative dredging developments. Again the benefits brought by dredging were felt throughout the community.

4.1.1 Later developments

UK dredging remained largely unchanged until the 1950s. The steam-powered bucket ladder dredger, larger, more powerful and able to dredge deeper than the machines of the nineteenth century, was still supreme although the stationary suction or sand pump dredger also achieved formidable outputs. But change was coming. The bucket ladder dredger with its attendant barges to carry away the dredged material was an expensive item of plant to operate. Many of the dredgers were very old – a 50 year working life was not uncommon – and in need of replacement. More efficient types of dredger were also now available. Thus for port maintenance dredging the bucket ladder dredger gradually gave way in the 1950s and 1960s to first the self-propelled grab hopper dredger and then the trailer suction hopper dredger. The former was not a novel type of dredger in the UK and both single and multiple crane versions could be found operating in many of the country's harbours by the close of the nineteenth century. But the introduction of diesel-driven dredging cranes, and diesel propulsion, reduced the operating cost and increased the efficiency. Large numbers of self-propelled grab hopper dredgers were built for the British Transport Docks Board which, following railway nationalisation, had become the largest port and harbour authority in the UK.

While suited to dredging within docks and alongside quays, the grab dredger was not ideal for approach channels. In these areas the trailer suction hopper dredger was more efficient. Highly manoeuvrable, able to operate in exposed conditions and not requiring the moorings and anchors needed by other types, the trailer dredger steadily replaced the remaining bucket dredgers. By 1992 the last bucket ladder dredger had gone from Britain's seaports and all maintenance dredging was being undertaken by grab or trailer dredgers, or in some cases such as Bristol, by specialist suction dredgers.

Another recent development, used at Harwich, is Water Injection Dredging. This type of plant injects water into the seabed and creates a fluidised layer which flows ahead of the dredger as it moves. It works best on slopes, such as riverside berths and dock entrances where the river channel is deeper than the berth. At the time of writing it is not regulated in the way that most dredging operations are, because the sediment is not removed above the water line or taken into a vessel.

The increasing cost of modern dredgers has resulted in further changes. With privatisation of the port industry, capital investment and its returns have to be carefully considered. Does a port invest say £20 million in a new dredger, or is the investment more profitably made in a revenue earning activity such as a new quay or improved cargo handling equipment? Where the real dredging requirement needs the constant use of a dredger, the capital investment may be justified. Associated British Ports, the privatised successor to BTDB, purchased both new and second-hand trailer dredgers for use in its Humber, South Wales and other ports. This later developed into the setting up of UK Dredging which now competes for contracts in other ports, and not only in the UK. Where the requirement is not continuous the alternative of employing a specialist dredging contractor makes greater economic sense. Thus ports such as the Tyne, Sunderland, Manchester, Thames, Belfast and the Clyde, all of which formerly operated major fleets of dredgers, have now disposed of their own craft and gone out to contract. This change has resulted in the majority of maintenance dredging projects in recent years being undertaken by contractors, although the majority of maintenance dredging quantity is still undertaken by in-house dredgers. The dredging undertaken by Forth Ports, Tees & Hartlepool, Associated British Ports and Mersey Docks using in-house dredgers accounted for around two-thirds of the total UK maintenance dredging activity in 1995. However, the general trend towards contract dredging continues, and in Scotland, the largest dredging operation Forth Ports, now uses contract dredgers.

4.2 TYPES OF DREDGING IN THE UK

4.2.1 Dredging for Ports and Harbours

4.2.1.1 *Environmental legislation*

Statutory control of dredging and disposal of dredged material in the UK takes place through a number of different pieces of legislation.

The disposal of dredged material in UK waters is licensed under Part 2 of the Food and Environment Protection Act 1985, (FEPA). In assessing whether a licence should be issued, the licensing authorities take account of the principles of the various international conventions on marine disposal, particularly the London Convention of 1972 (LC72) and the OSPAR Convention. The licensing authorities for England and Wales are the Department for Environment, Fisheries and Rural Affairs, (Defra), for Scotland the Scottish Executive for Rural Affairs Department, (SERAD) and for Northern Ireland the Department of the Environment Northern Ireland, (DOENI).

The dredging activity itself may be specifically authorised under a local Harbour Act but if not has to be approved by the Department for Transport, DfT, under the Coast Protection Act 1949 and sections of the Merchant Shipping Act. The Department may also enforce the requirements of Statutory Instrument No 124, Harbour Docks and Piers and Ferries – The Harbour Works (Assessment of Environmental Effects No 2) Regulations 1989 and Statutory Instrument No 3445 Harbours, Docks, Piers and Ferries, The Harbour Works (Environmental Impact Assessment Regulations) 1999.

In addition to this, European Directives such as the Environmental Impact Assessment Directive (97/11/EEC), Habitats Directive (92/43/EEC) and the Birds Directive (79/409/EEC) must be taken into consideration by the appropriate licensing authority. If dredging or disposal is likely to affect the conservation value of a European site, occurring either outside or within such an area, the licensing authority must ensure that an appropriate assessment is carried out.

A review was undertaken by DfT in 2003 of the way in which a range of activities are consented in the marine environment, and this may ultimately result in changes to the way in which marine dredging and disposal are regulated in the UK.

4.2.1.2 *Capital dredging*

In the last 20 years there have been a number of capital dredging projects in the UK. Since the UK is already established as a maritime trading nation, with more than 150 ports, most of the capital dredging projects have been to extend existing facilities and to deepen berths or access channels, rather than to create new ports.

Capital dredging projects have their own particular requirements. In addition to the legislative controls (see section 2.1), many other factors may constrain the operation. The availability of funding to resource the development is a major consideration, since most UK ports are under private ownership, and both capital dredging and maintenance are at the expense of the port owners and operators. The availability of suitable dredging plant, in an increasing global market, is a consideration, as is the availability of suitable disposal sites for the potentially large volumes of dredged material. Increasingly developers attempt to use some of the dredged material within the construction elements of a project to maximise the re-use of suitable material, and to reduce quantities that require off site disposal. Virtually all capital dredged material from the UK which cannot be used beneficially either within the project, or nearby, is disposed of to sea in sites designated for the purpose.

Some examples of recent capital dredging projects in the UK are given below.

Capital dredging of Felixstowe channel and Harwich Haven approaches off the east coast of England

During the 1990s the Trinity berths at Felixstowe were extended to accommodate larger container ships, to maintain the port's competitive position. At this time Felixstowe channel was deepened once in 1994 to 12.5 m below chart datum, CD, and later in 1998/2000 to −14.5 m CD.

In this latter operation the area dredged extended from the northern end of the Trinity berths at Felixstowe to the deep water at the south of the Shipwash buoy in the outer channel, a distance of some 27 km. Over the length of the channel a variety of sediments were required to be dredged, including sand, gravel, rock, and clay. The Harwich Haven Authority who as the Navigation and Conservation Authority oversaw the dredging and disposal process, made extensive efforts to make use of as much of this material as possible. Much of the sand and gravel was used within the Port of Felixstowe, and other beneficial use schemes accounted for the remaining sand and a small proportion of fine material. There still remained about 15 million cubic metres of stiff clay for disposal. This material was deposited at "Roughs Tower", a deposit ground in the Outer Thames estuary. This site had been used for many years for the disposal of both maintenance and capital dredged material. However, following the development of a crustacean fishery nearby and concerns over the dispersive capacity of the local environment in response to such large quantities of material, this deposit was accompanied by effective closure of the site in 2000.

The disposal operation was conducted in such a way as to promote containment of the material within the licensed boundary. This involved the initial construction of a bund along the northern and western edges of the disposal area, employing cohesive clays and rocks, followed by sequential infilling of central areas, including the deposit of softer and more heterogenous material. Gravel was sprinkled over the deposited material on the western side of the area, with the aim of enhancing the habitat for colonisation by commercial shellfish. Subsequent monitoring of the deposit site has provided clear evidence of the continued integrity of the bulk of the recently deposited material, some 14 months after cessation of disposal. This suggests that a new habitat suitable for re-colonisation by bottom dwelling animals, including commercial shellfish species may indeed have been formed. Monitor the site is continuing under Defra's annual disposal site monitoring programme.

Southampton Port

The most significant capital dredging operation in recent years in the Solent area on the south coast of England was conducted by Associated British Ports, Southampton during 1996 and 1997. The deepening of the navigation channel to Southampton Port resulted in the removal of approximately 7 million cubic metres of fine sands and silts. The majority of these were deposited at the designated disposal site at Nab Tower. This site is used by many operators and regularly receives both maintenance and capital dredged material from the many wharves and marinas in and around Southampton water, other local rivers and harbours and the Isle of Wight coast. The historic use of this site for disposal of dredged material and the encroachment of the activities of marine aggregate extraction industry into the locality of the site has meant that the site has received extensive monitoring attention.

A proposed extension of the ABP Dibden Bay terminal, if approved, will give rise to large volumes of capital dredge material during its construction phase. While beneficial use of some of the dredged material will be sought, including use within the project, it is likely that a proportion of material will be disposed of at the Nab Tower site.

On completion of the last channel deepening the offshore disposal site at the Roughs Tower was closed for all disposal activities and a programme of habitat improvement and monitoring initiated. A new dispersive offshore disposal site for maintenance dredged material at the Inner Gabbard considerably further offshore was opened in 1998. Presently there is no offshore disposal site for capital dredged material although a new site is being sought.

Barrow-in-Furness
A major dredging programme to deepen the approach channel to Barrow-in-Furness, on England's northwest coast was undertaken in 1991. Barrow-in-Furness is a traditional centre of shipbuilding expertise, and a decision to construct Trident nuclear submarines at Barrow brought with it a need to deepen the access channel to allow the submarines to exit the port.

The cutter suction dredger Leonardo da Vinci, at the time the world's largest dredger, was one of the vessels used to deepen the channel. In all approximately 8 million tonnes of dredged material were removed for disposal.

There was no existing marine disposal site in the locality suitable for this quantity of capital material and so, after extensive investigations, a new site, Barrow "D" was designated to the south of Walney Island. This site was chosen to avoid interference with other uses and users of the area including fishing and navigation. Careful placing of the dredged material in the site was necessary to ensure that excess shallowing did not take place, to avoid navigation hazard.

Aberdeen, Scotland
Capital dredging work was undertaken in Aberdeen Harbour at four locations to deepen and widen berthing areas during 2002. A mixture of sand, gravel and clay was removed for disposal at the offshore site.

Aberdeen Harbour has been in existence for over 850 years and dredging has been done throughout its history. For at least the past 50 years a combination of grab, suction, back hoe and bucket dredgers have been used successfully and the material placed in the same offshore site. Aberdeen is the largest support harbour for the North Sea Oil industry.

Aberdeen Harbour has been built on the former delta of the mouth of the River Dee. The harbour and the River Dee are connected to the North Sea by a navigation channel. Both the harbour and channel are susceptible to continued progressive natural infilling from both river-borne silts and muds and sea-borne sands. Sediment transport is due to a complex action of tides, currents and wave action and consequently cannot be controlled. The River Dee is a Special Area of Conservation under the Habitats Regulations.

Port of Belfast, Northern Ireland
The Port of Belfast carried out major capital dredging in 2001 to accommodate an ever increasing demand for the port and to improve safe passage of vessels entering into the port. The disposal project accounted for almost 4 million tonnes of dredged material to be disposed of at sea and a further 0.5 million tonnes for land reclamation. As Belfast Lough

has many environmental designations, (Area of Special Scientific Interest, Special Protection Area and Ramsar site) and marine cultural and a substantial commercial aquaculture interests, an extensive licence application appraisal was necessary.

This resulted in a number of specific monitoring conditions being included in the Port's sea disposal licence such as:

- Pre and post video surveys of the disposal site;
- Acoustic Doppler Current Profiling (ACDP) surveys for definition of sediment plumes during dredging and disposal operations;
- Turbidity monitoring at the dredging site;
- ROXANN survey (hydro acoustic method of discriminating between seabed material types) of inner Belfast Lough to assess the extent of commercial mussel beds pre, during and post dredging activity.

4.2.1.3 *Maintenance dredging*

The United Kingdom, consisting of England, Scotland, Wales and Northern Ireland is surrounded by water and has many ports on its coastline and in its estuaries. A few may be described as natural and require little maintenance but in most cases the coastline or estuary has been modified. It is a fact that most ports or channels in the UK created by capital dredging require ongoing maintenance dredging. There are some exceptions where natural deep water exists such as Milford Haven on the west coast of Wales, and some of the west coast of Scotland fishing ports.

Maintenance dredging could be simply regarded as a routine job just like sweeping the house to keep it clean but is it so simple? Where does the sediment come from? Where would it have gone if it had not ended up in your dredged channel or harbour? What does that tell us about where we should put it after we have dredged it?

For the ports that have to carry out substantial maintenance dredging each year most of the work is done using trailing suction hopper dredgers. Small ports tend to use grabs or backhoes working with barges. In previous years many ports, even some of the small ones would own and operate their own dredging plant. In some case this would consist of a whole fleet of dredgers. Today there are few that do so, most preferring to let single or term contracts. Associated British Ports, which owns a number of ports around the UK, recently set up its dredging fleet operation as a separate company, UK Dredging, which now competes on the international market, and currently has the contract to dredge Grangemouth, the largest maintenance dredging project in Scotland.

The majority of the UK's maintenance dredged material, most of it mud, is placed at sea disposal sites which today have to be licensed. Records are kept by the Centre for Environment Fisheries and Aquaculture Science (CEFAS) of the amount placed at each site each year, its source, whether the material is capital or maintenance dredgings and some information about its contaminant content. The statistics thus give an indication of the amount dredged from each region and the major ports and harbours. The statistics for the year 2000 have been used to illustrate the amounts from each in the following discussion. The total for 2000 was about 16.5 million tonnes dry weight. However, there are variations from year to year both at individual sites and in totals each year as is illustrated in Figure 4.1.

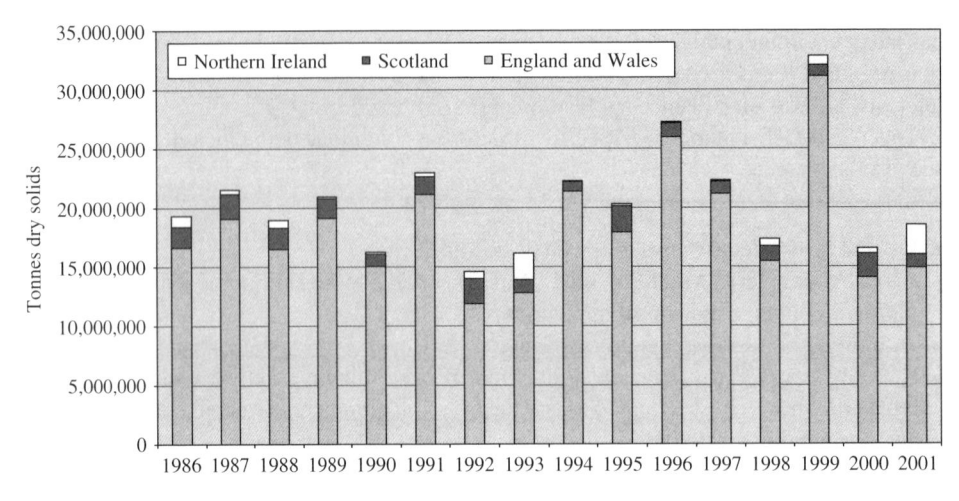

Figure 4.1. Quantities placed a licensed disposal site 1986–2001.

Some of the variation above is due to capital projects, which have not been separated out in the statistics.

In 2000 there were 132 licensed dredged material disposal sites of which 32 were not used in that year. 29 sites received more than 100,000 tonnes of which 6 sites exceeded 500,000 tonnes and a further 4 sites exceeded 1million tonnes. The figures given in the above diagram are dry tonnes. Sometimes figures are given in "wet tonnes" ie measured by displacement of the vessel carrying the material to the disposal ground.

In the following paragraphs we take a tour of the major regions around the UK coastline, finding that they classify quite readily into the types of problems that they have to deal with. The list is not comprehensive and more detail is given of some than others to illustrate the issues by example.

Clearly most dredging is carried out in England and Wales. Most of the dredging in Scotland is needed on the east coast in the Firth of Forth where the tour begins. Maintenance dredging is required at Leith, Grangemouth and the naval port of Rosyth accounting for 1.2 million tonnes of the total UK figure of 16.5 million tonnes (ie 7%).

Grangemouth Dock entrance is accessed by a channel through the Forth Estuary. This channel is stable over most of its length but is dredged for the final 1.5 nautical miles, removing approximately 1.5 million wet tonnes per year. It opens into a turning area and bellmouth entrance outside the western approach to the docks and is protected by open piled lead-in jetties. Grange Dock is connected to the dock entrance via the Eastern Channel and lock. The bellmouth lies immediately west of the Kinneil mudflats and east of the Skinflats mudflats. The action of the waves on the mudflats, combined with turbulence created by the movement of the tide against the west lead-in jetty, results in the movement of suspended solids from the mudflats into the bellmouth area. This causes the bellmouth area to silt up and, if the maintenance dredging did not take place, the bellmouth would silt up at a rate of about 1 m per month and would rapidly become un-navigable.

Continuing the tour down the east coast of England the first major ports are on the estuaries of the Tyne, Wear and Tees. In 2000 these had dredging quantities (rounded) of about 1 million dry tonnes of which half was from the Tees. This reflects the fact that the Tyne and the Wear have declined while the Tees port continues to do well. A detailed description of the Tees dredging is given as representative of the region.

Tees estuary
It is a common misconception that the need for maintenance dredging in the UK is caused by sediment being carried down the rivers. In the Tees estuary the quantity of sediment carried in by the Tees river itself is insignificant when compared to the maintenance dredging rate which is accounted for by sediment entering from the North Sea.

Historically it was a shallow meandering estuary with large intertidal areas (HR Wallingford, 2002). The intertidal areas have been reduced from about 3,300 ha to less than 500 ha over the period 1740 to 1974. The effect has been to reduce the tidal volume (or prism) of the estuary which, combined with deepening of the entrance area, has reduced the velocities in the estuary to the point where they are not enough to re-suspend deposited sediment. It thus acts a sink for sediment of marine origin, stirred up by storms in the North Sea and transported into the estuary by the flood tides. Once in the estuary it settles in the relatively quiescent conditions prevailing there.

Between 1895 and 1923 the rate of dredging remained essentially the same, between 400,000–500,000 yds^3 (in situ measurement). After 1923 the indications are that the rate of dredging increased substantially, almost double the previous rate, until the war years when a decline took place. After the war the lower dredging rate persisted until 1952 when it was restored to the immediate pre-war rate where it remained until at least up to 1960.

Siltation rates are not necessarily the same as dredging rates because the latter are affected by availability of plant, budget control etc. However, unless there is a trend for net loss of depth or increase in depth the siltation and dredging rates must average out to the same figures. In fact the analysis carried out in 1986 and updated in 1991 by HR Wallingford showed that the average siltation rate was about 1.5 Mm^3/year. Prior to 1986 the dredging rate had exceeded the siltation rate by an average of 11% resulting in a net deepening of the estuary. Since that time dredging has fallen below expected siltation rates and this means that the average bed levels must be rising again. The present dredging rate of below 1 Mm^3/year represents a significant shortfall. (Note that the units in Figure 4.1 above are dry tonnes and not m^3. A very approximate conversion factor of 2 is applicable.)

Historically the Tees dredged material was placed on land. For economic reasons it was placed as near to the point of dredging as was allowed. It thus contributed to the loss of intertidal area, but at the same time created habitat that now has special conservation status, perhaps one of the ironies of this conservation conscious age.

The present disposal practice is to take the material out to the licensed site at sea. This is a dispersive site and the evidence is that the deposited material is transported in the general direction of the coastal drift southwards. It can be argued that this is the best thing to do with it because it puts the sediment back approximately where it came from and thus maintains the natural coastal drift.

Humber
Moving south down the east coast of England the Humber estuary is home to a number of ports such as Grimsby, Hull and Immingham. In 1974 the British Transport Docks Board reported that almost no dredging was required to maintain the navigation channel in the

estuary. An approximate long-term average sediment budget indicates that there is a net sedimentation of 170 t/tide or about 0.1 Mm^3/year, which is only a small fraction of the flood and ebb transport at the mouth. The volume of the estuary is increasing by about 0.4 Mm^3/year so it can be argued that there is a deficit in sediment supply so dredging is only needed to keep the channels themselves open and there is no need, it might even be regarded as undesirable, to remove sediment from the estuary.

However, dredging is certainly required today for deeper draught vessels. The average for 1985–1993 was about 1.4 M tonnes (dry weight)/year. The figure for 2000 was about 4 million tonnes but this included some capital dredging. This material was deposited at sites within the estuary to avoid sediment removal from the system.

Orwell and Stour
The major port of Felixstowe is just inside the entrance of the combined estuarine mouth of these two rivers. Again the source of sediment is predominantly marine. Routine maintenance is needed to maintain the deepened manoeuvring and berthing areas. The approach channel is largely maintenance free. Presently approximately 1.5 million dry tonnes of material are removed each year to maintain access to the port. Since the last channel deepening (1998/2000) a proportion of the maintenance dredged material has been returned back into the estuary system through a sediment replacement programme rather than being placed offshore. The aim of the sediment replacement programme is to prevent an acceleration of the underlying rates of erosion of intertidal areas in the Stour and Orwell estuaries.

Thames estuary
Next down the east coast is the Thames estuary, where, as was the case with the Tees and the Humber, the river itself contributes very little sediment compared to the amount of tidally transported sediment. For example, at Woolwich approximately 100,000 tonnes of sediment moves up estuary with the flood tide and a similar quantity moves downriver with the ebb tide. A typical Thames river input is extremely small compared to this as a source of siltation. Similarly there is no significant input at the seaward end. The potential for siltation is thus the high suspended sediment load in the estuary due to the continual redistribution of sediment by the strong currents. Long term routine maintenance dredging of the channels to the dock areas near central London therefore resulted in a gradual lowering of the average bed level. Routine dredging of the upper reaches stopped in the 1980's as the need for depth in the upriver docks declined and attention was focused on maintaining the riverside berths and dock entrances. This resulted in surprisingly little loss of depths in the channels but a gradual rise in the intertidal bank levels.

Dredging in the Thames Estuary in 2000 amounted to less than 40,000 tonnes.

South east coast ferry ports
Moving south of the Thames estuary and along the Kent and Sussex coast the character of the ports changes from predominantly estuarine to coastal harbours ie Ramsgate, Dover and Folkstone. Relative to the estuarine ports these require little maintenance dredging, the source of siltation being mainly the re-suspension of seabed material by waves and transport by tidal currents. Ramsgate and Dover are believed to be near enough to receive some suspended sediment from the Thames estuary, though the concentrations are small.

Southampton water
Southampton water is gradually deepening due to tilting of the tectonic plate. Due to the unusual tidal split around the Isle of Wight it has double tides which means it has naturally deep water for more of the time during the tidal cycle. This area includes Portsmouth from which a number of ferries run as well as naval vessels. Disposal of dredged material for this area in 2000 amounted to just over 0.5 million tonnes although this included some capital dredging.

The south west Penisula
The sea approaches to Poole Harbour require periodic maintenance though the harbour itself in a natural harbour requiring little dredging. This is one of the few examples where beneficial uses of dredged material have been put into practice on a large scale, namely to use the sandy material arising from capital and maintenance dredging to recharge the eroding beach at Bournemouth and Poole. The material being clean and of similar character the only problems were the political and logistical ones, and the need to avoid interference with others users during the replenishment process.

Plymouth, the well known naval port where Sir Walter Raleigh insisted on finishing his game of bowls before defeating the Spanish Armada is at the mouth of the river Tamar. It requires little routine maintenance, although there is periodic dredging of some of the naval docks.

The next significant port is Falmouth though today it is better known for its marinas, fishing and small boat activity. Maintenance dredging quantities are small.

The Severn estuary
The next large ports as we continue the tour round the coast are those on the Severn estuary. The south side is English and the north side is Welsh. The Severn is characterised by its very high range tides, the second largest in the world, having amplitudes up to about 14 m. This in turn gives rise to its other dominating characteristics of very high flow velocities (up to 6 or 7 knots) and very high suspended solids concentrations (in excess of 100,000 ppm on spring tides). Not surprisingly the ports tend to suffer from high siltation rates. The actual dredging rate is bound to be higher than is discernible from the licensed disposal figures because some of the ports practice various forms of agitation dredging (eg Avonmouth), discharge via pipelines and flushing (Sharpness). The ports on the Welsh coast, Port Talbot, Swansea, Cardiff and Newport accounted for some 1.5 million tonnes of disposed material in 2000.

West coast of Wales
Milford Haven is blessed with naturally deep water able to take supertankers without the need for large quantities of maintenance dredging.

Fishguard serves the Irish ferry trade but again requires little maintenance.

Further north, Aberystwyth today is essentially a marina that dries out at low water.

At the north west extremity the port of Holyhead on the island of Anglesey was responsible for about 100,000 tonnes of dredged material in 2000.

Dee and Mersey
Travelling east along the north Wales coast the first major inlet is that of the River Dee, which also marks the border with England. At one time this was a much more significant port area

than it is today, though the port of Mostyn is still active. It requires maintenance dredging of 200,000–500,000 m^3/year. A recent capital dredging programme there has resulted in an increased need for maintenance dredging.

The Mersey is home to the ports of Liverpool and Birkenhead. The total dredging disposal amounted to about 1.2 million tonnes in 2000. The main source of sediment is believed to the southerly movement of sediment eroded from the coast in the area of Formby.

The Mersey also provides access to the Manchester Ship Canal. Approximately 0.5 million tonnes are removed each year from the canal and placed on land in lagoons that are eventually restored as agricultural land.

4.2.2 Dredging for beach nourishment

Beach recharge (also known as beach nourishment, replenishment or feeding) means supplementing the natural volume of sediment on a beach using material from elsewhere. The "imported" material may be either sand or shingle. The use of beach recharge has been steadily growing in the UK, as in many other countries, over the last 25 years, and this is largely because of the development of dredging and delivery systems that make the operations cheaper, in real terms, as the years pass. The annual amount used in this way is currently about 2.2 million tonnes (ie about 10% of the total dredged) (Harrison, 2003).

The provision of a wider, higher beach is usually regarded as a benefit by seaside towns, as it will provide a better amenity for residents and visitors alike.

Unlike seawalls and revetments, beaches are dynamic features being subject to the mobilising influence of waves and tides. To provide an effective defence, a beach must maintain a required volume (or cross-sectional area) at every section along the frontage.

In the long term, losses of the recharge material and unwanted changes in the beach shape can be managed in several ways, usually by a combination of adding extra material as and where required and installation of beach control structures. Without such measures, the expected lifetime of a beach recharge scheme may be short, although this will greatly depend on the particular site. In suitable locations, a beach recharge scheme can be expected to remain effective for 5 to 10 years before any major addition of further sediment.

A useful account of beach nourishment projects in Europe is given in Hanson et al. (2002).

The material for beach recharge is usually obtained from the seabed, from areas licensed for the commercial production of aggregate for the construction industry. In recent years, there has been a growing trend towards using sediment dredged from navigation channels to ports, and from areas of the seabed licensed solely for supplying beach recharge material.

The first substantial beach recharges in the UK, at Portobello, Edinburgh in 1972 and at Bournemouth in 1973/74, were carried out using sand dredged from the offshore seabed with the sand being pumped ashore through a pipeline. The scale of such operations has increased in recent years with much larger sand recharges carried out between Happisburgh and Winterton, in Norfolk and between Mablethorpe and Skegness, in Lincolnshire between 1997 and 1999.

Shingle recharge schemes are usually designed on the basis of average particle diameters of between 5 mm and 15 mm. This originally made delivery by pipeline impractical and early recharge schemes involved delivery of material by bottom-opening barges moored in the inter-tidal zone. In more recent schemes, notably at Seaford in Sussex in 1987 and at

Hurst Castle Spit, Hampshire in 1996, improvements in equipment have allowed delivery by floating pipelines, in the same way as for sand.

More rarely, beaches have been formed using much coarser material (50 mm to 500 mm); examples of recharged "rock beaches" can be found at Hilton Bay near Berwick-on-Tweed, and near Prestatyn in North Wales.

4.2.3 Dredging for aggregates for construction

The UK marine aggregate industry is second in scale only to Japan. Over 60% of the 22–26 million tonnes dredged annually supplies the construction industries of South East England and the coastal strip of north-eastern France, Belgium and The Netherlands. Much of this sand and gravel is extracted from licenses in the southern North Sea off eastern England. Other important areas are the Bristol Channel and the Irish Sea where sand is supplied without the need for onshore processing (Seaman, 2003).

The main differences between marine and terrestrial deposits are shell and sea salt content, the former being low due to the terrestrial origin of most deposits and the latter reduced to acceptable levels (where necessary) by shore-side processing. Marine aggregates are geologically more "far travelled" than land based deposits, commonly resulting in a more rounded coarse fraction and cleaner, well-graded sand fraction, enhancing concrete workability and strength with a low water/cement ratio in the mix. Marine aggregates are therefore used in many high specification construction projects.

In order to be economical, the industry prospects for sites that will yield material of a suitable quality (with a minimum of processing) for sale either in the UK or other European countries. This means that operations are typically restricted to a particular type of environment, such as one in which the water depth is not too great and the bed material is sand or gravel with a minimum amount of fine material. In most cases the material is processed on board by washing out unwanted fine material and discharging it overboard (screening). Clearly this process can be a significant generator of plumes of suspended solids, depending on the fines content of the aggregates. Such dredging often takes place in high-energy conditions (ie. environments characterised by strong currents and/or waves).

Aggregate dredging for sands and gravels mainly uses trailing suction hopper dredgers for which overflow, especially that generated by screening, is acknowledged to be the dominant source of release of fine sediment (Bray et al., 1997).

Over the past decade the UK marine aggregate dredging industry has recovered an average of 24 million tonnes per annum (BMAPA website). Just over half of this resource is landed on the south and east coasts, representing a third of the UK construction sand and gravel requirements. Marine aggregates are, however, becoming more important due to increasing constraints on land-based quarrying, and the UK government is promoting greater use of marine resources.

There are 70 licensed areas spread over 7 regions. The total licensed area represents less than 1% of the seabed. The area actually dredged is much less. On average only about 15% of the licensed area is dredged in any one year, much of it at low intensity.

Half of the aggregate dredged around the British Coast goes to construction, 20% to coastal defence and about 30% is exported to near neighbours in Europe. The amount produced is the equivalent of 50 medium sized gravel pits on land. One 5,000 t hopper dredger transports the same amount as 250 trucks.

Some sites have been used now for 30 years and are becoming depleted, so the industry has to prospect for new sites.

Better control has resulted in a progressive reduction since 1998 in the area dredged so that by 2001, 90% of dredging took place in only 13 km^2 of the sea bed.

Most dredging takes place within the 12 mile limit, which is also the area most used for fishing, navigation and recreation. To minimise interference the industry has co-operated with Crown Estates to produce Regional Zoning Plans to define which areas will be dredged at any one time (The Active Dredged Area). This information is widely distributed and is updated twice each year.

4.3 ENVIRONMENTAL ISSUES

Environmental issues and the impact of the dredging and disposal operation on other users of the dredging and disposal areas are important considerations when permissions for dredging and disposal are sought. The move towards spatial planning at the coast, and towards Integrated Coastal Zone Management, ICZM, will mean that ultimately dredging and disposal projects will be considered alongside the very many other activities at the coast, and consideration will be given to the cumulative environmental impact of a wide range of activities. A number of potential environmental impacts of dredging and disposal activities are currently taken into account in the assessment process.

4.3.1 Dredging

During the dredging operation there is potential for direct interference with other users of the area, for example, for navigation, fishing, conservation or amenity. The authorities permitting the dredging require account to be taken of such interference, and the operation to be planned to minimise it. There is some evidence of potential disturbance of fish from the noise generated during aggregate dredging, but the main environmental effects of a maintenance dredging operation are likely to relate to sediment disturbance, and the resulting turbidity and other water quality effects in the water column. If the sediment has high concentrations of contaminants associated with it, there is potential for contaminant redistribution, and sometimes contaminant release from the sediment into the water where it is more available for up-take by living organisms.

The choice of dredger influences the potential loss of sediment during the operation, and in some sensitive environments a dredger that has relatively low loss of sediment is required (Burt et al., 2000). Other control measures such as silt curtains exist but have not been used extensively in the UK. Efforts to minimise sediment loss or movement from the dredged area would be applicable in areas close to shellfisheries, or eel grass beds for example, where sediment settling from the water column can have adverse effects. Not all dredged areas are so sensitive to sedimentation, and assessment of the potential impact, and consideration of the type of dredging equipment is made on a case-by-case basis. In some instances dredging has been restricted to certain times of year, so called "environmental window". An example of a use of an "environmental window" occurred in the licence for a capital dredge at Holyhead in Wales in 2002, when blasting of rock was not allowed during the months of April to July inclusive to minimise the disturbance to breeding birds.

The dredging of the River Lagan, which feeds into Belfast Lough, is an example where the dredging could only be permitted in a "small" window of opportunity. The River Lagan, through intensive water quality improvement and fish stock enhancement initiatives has become an established salmonid river. In order to protect salmonid migration and prevent fish mortalities through increased suspended solids and subsequent drops in dissolved oxygen (DO), time restrictions were placed on the dredging operators when dredging could not occur (no dredging between mid May and September). In addition to this DO monitoring in the vicinity of the dredging, was listed as a condition of the sea disposal licence. Where drops of oxygen were detected near the months of May and September dredging activity was halted. No detrimental impacts on salmonid migration were observed during or after dredging activity.

Long term effects on the marine environment resulting from morphological changes due to the dredge and subsequent changes to the hydraulic regime can include wave refraction and reflection effects, the alteration of current and sediment transport paths and changes to siltation patterns. At Harwich on the east coast of England, the creation and subsequent deepening of the main access channel to the Port of Felixstowe was predicted to reduce the sediment supply into the Orwell and Stour estuaries. Here, licence conditions were placed on subsequent dredging programmes to disperse some dredged material into the estuaries upstream of the Port to increase the supply of upstream sediment required for maintaining intertidal areas.

4.3.2 Disposal

Disposal of dredged material to land is relatively limited in the UK. Potential environmental impacts arise from the creation of sediment lagoons, in that loss of valuable land can occur, but conversely there are examples in the Thames Estuary where old sediment lagoons have become valued conservation areas. The Port of London Authority is examining with conservation authorities if the suitable management of dredged material lagoons can bring about environmental benefit, rather than loss. The example of the Tees dredging was mentioned earlier where the disposal site became a nature reserve.

The majority of dredged material in the UK is deposited at sea. Sites designated in the last 30 years or so, since national disposal legislation was first enacted, (The Dumping at Sea Act, 1974) have been chosen to be in areas where environmental sensitivity and the potential for interference with other uses of the area is relatively low. Nevertheless, the seas around the UK are very busy, with extensive conservation interests, fishing activity, navigation, pipelines, cables, extraction activities, archaeological and amenity interests, and hence the selection of sites for disposal of dredged material is difficult. As a result many of the recently designated disposal sites are many kilometres offshore, and may be distant from the dredging site.

Sea disposal licences usually carry conditions designed to minimise any adverse effects on the marine environment. Licences issued for the Hurst Fort deposit site to the west of the Isle of Wight, for example, carry conditions requiring disposal only on the ebb-tide, to ensure the initial dispersion of sediment is away from the sensitive shellfishery areas to the east.

Contamination of dredged material by heavy metals, antifouling agents, oil and pesticide residues, and other chemicals derived from industrial, dockside or farming practices can

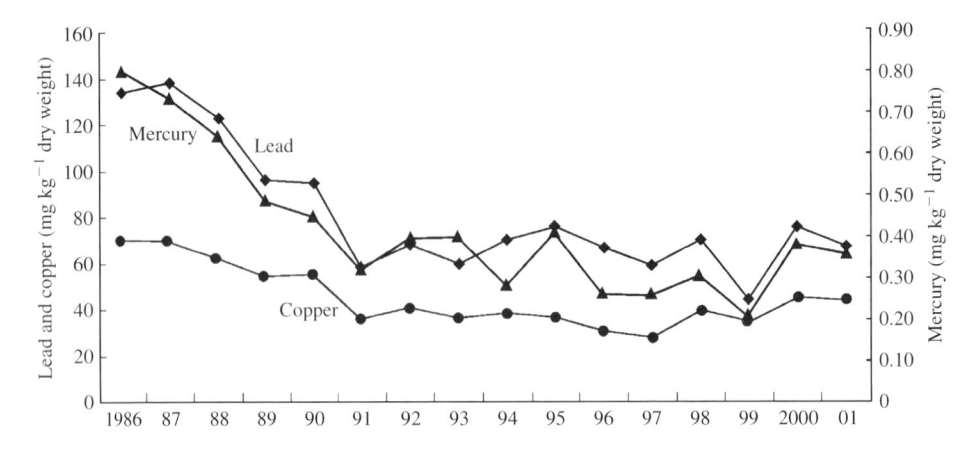

Figure 4.2. Heavy metal concentrations in dredged material 1986–2001.

impose a constraint on their disposal. National legislation, and International Conventions require that the chemical contaminant content of dredged material be assessed before sea disposal permit can be given. In the UK, each application for a dredged material disposal licences is assessed, and a decision made using a "weight of evidence" approach. This means that the quantity and chemical quality of the dredged material, and its potential for biological harm are taken into account in the decision on licence issue, also taking into account the characteristics of the disposal site. Most UK dredged material is analysed for a range of contaminants during this assessment process. Licences are not issued for the sea disposal of highly contaminated material. Some contaminants give rise to particular concern, for example organotin antifouling residues, (tributyl tin, TBT and dibutyl tin, DBT) which are present in many dock sediments. The presence of TBT, or some other metal contaminants particularly mercury, and oil residues, has been the reason for most of the refusals of sea disposal licences in England and Wales (Murray, 1994).

The most effective way of dealing with this contamination problem in the long term is to reduce the contamination reaching the sediments. Better dock-side practices have their part to play, but the reduction of contaminants at source is essential for a long term solution. The reduction of some contaminants entering the UK's river and marine environment from industrial sources, as a result of improvements in industrial processes and improved waste treatment practices has had a dramatic effect on the quality of UK dredged material disposed to sea. The average concentration of some metal contaminants in UK dredged sediments analysed prior to disposal to sea over the years 1986 to 2001 is shown in figures above.

Reductions in other contaminants of particular international concern may be a requirement in the future if continued sea disposal of dredged material is to be assured.

4.3.3 Beneficial use

In the UK there is increasing emphasis on the use of dredged material, to consider it a resource rather than a waste material from the dredging process. In assessing licence applications for sea disposal, the UK authorities require that possible use of the material has been given rigorous consideration. In some circumstances, dredging techniques may require adaptation to obtain the material in a form suitable for use. When dredging forms part of a larger project, consideration is given to the reuse of the dredged material within the project, for example as hydraulic fill. Few disposal licences for substantial quantities of coarse dredged material have been granted in recent years as a result of the emphasis on reuse.

The beneficial use of fine grained dredged material, such as maintenance silts and estuarial muds which represent the majority of dredged material in the UK is more difficult. Use on land is limited due to their poor engineering properties and the time taken to dewater, although techniques exist to produce top-soil and other products in appropriate circumstances. Within the UK there has been increased emphasis on the relocation of fine-grained dredged material in such a way as to derive environmental benefit. As a result a number of beneficial use options have developed whereby the material is used to recharge or recreate intertidal habitats. Dredged material has been shown to successfully create new mudflats and saltmarshes that eventually become capable of functioning like natural systems, with associated nature conservation and flood protection benefits. Up to now, the beneficial placement of maintenance dredged material has been limited to small or medium scale trails, as research is being undertaken to increase our understanding of recovery processes (Fletcher et al., 2001).

Between 1994 and 2002, approximately 12 beneficial use schemes using fine grained material were set up.

The success of beneficial use schemes in the last 8–10 years has significantly contributed to an increasing willingness to identify methods of placing fine dredged material other than the traditional foreshore replacement. One of the main drivers was the assurance of the regulators and nature conservation bodies that the retaining of fine dredged material within an estuarine system is one of the more appropriate applications for beneficial use. The ongoing research is likely to provide important guidance of the best ways of achieving benefit from the deposits.

4.4 FORWARD LOOK

Pier Vellinga, professor at the University of Amsterdam's Institute for Environmental Studies presented a fascinating and challenging analysis of dredging in a changing environment at Dredging Days 2001 (Vellinga, 2001*). He addressed, at a fundamental level, our interaction with the environment from a social and cultural perspective. I quote *"How long can a species [mankind] grow exponentially in numbers, in particular when such growth comes with increasing loss of other species? This question together with the history of being very vulnerable to the forces of nature makes our species presently rather insecure in dealing with nature. It should not be surprising therefore that political decision making regarding the management, use and protection of nature is often seen as erratic and irrational to the actors involved."*

With this backdrop he examined three main views of nature, the **conservation view** (conserving and restoring according to some historical reference situation "when everything was good, or much better than it is today!)", the **development view** (protection of existing nature and development of new natural sites) and the **functional view** (the value of nature

---

* A revised text of this paper is included at the end of this book.

is primarily related to the welfare it provides for society [ie. it only has value if we find it important]). How we view nature affects our decision making process.

Vellinga calls on us not to see development as being in opposition to conservation but to try to find creative ways of doing both at the same time, what he called "co-evolution", "how to interact with nature, not how to beat it". This requires a deeper understanding of the ecological processes.

Vellinga then showed how this interaction may bring to an end the ever increasing size of ships and consequent deepening of ports because the environmental effects would reach an unacceptable level. The consequent increasing need for more infrastructure, roads and rail in already congested and heavily populated areas might drive us to seriously reconsider the shipping trends and lead to perhaps more smaller ports in less congested areas. But the same considerations may lead to increasing use or refurbishment of inland waterways which many would argue is a more acceptable way of transporting bulk goods.

Whichever way the trends take us, with its many ports, rivers and exposed coastline, the UK will continue to have a requirement for substantial amounts of dredging. The wet weight of material dredged for port maintenance is about 5% of the total freight tonnage handled by the ports, which, in turn, generates the dredging requirement (Riddell, 1995). While the rationalisation of shipping accommodation in many of the traditional UK ports is now complete, privatisation has resulted in further close scrutiny of dredging need and costs. Thus the quantity of dredging undertaken may well fall with further increases in dredging efficiency and greater refinement in the method of measurement. Even so, dredging will always remain a substantial activity. It is quite simply **essential** for a country dependent upon maritime trade.

REFERENCES

Bray RN, Bates AD and Land JM (1997). Dredging – a handbook for engineers. Second Edition. Pub Arnold, London 1997.

Burt TN, Roberts W and Land J (2000). Assessment of sediment release during dredging: a new initiative called TASS. Proceedings of the Western Dredging Association Twentieth Technical Conference (WEDA XX) and Thirty-Second Annual Texas A&M Dredging Seminar (TAMU 32), Warwick, Rhode Island, USA June 25–28, 2000. Pub Texas A&M University, Texas 77843 – 3136 (CDS Report No 372).

Burt TN and Cruickshank IC (1999). Uses of recycled dredged and other materials in construction. HR Wallingford Report SR 555, July 1999.

Fletcher CA, Stevenson JR and Dearnaley MP (2001). The beneficial use of muddy dredged material. HR Wallingford report SR 579, April 2001.

Gower GL (1968). A history of dredging. Paper 1, Symposium on Dredging. Institution of Civil Engineers, London 1968.

Hanson H, Brampton A, Capobianco M, Dette HH, Hamm L, Laustrup C, Lechuga A and Spanhoff R (2002). Beach nourishment projects, practices and objective – a European overview. Proc. Coastal Engineer, Vol47, No2, Dec 2002; Special Issue "Shore Nourishment in Europe" pp 81–111. Pub. Elsevier ISSN 0378-3839.

Harrison DJ (2003). European overview of marine sand and gravel. Proc. EMSAGG Conference "European marine sand and gravel – shaping the future". February 2003, Delft, Netherlands.

Murray LA (1994). Environmental considerations in licensing procedures. Conference on dredged material disposal – problems and solutions, Institution of Civil Engineers, London 1994.

Riddell J (1995). Two hundred years of benefit: UK dredging 1795–1995. 14th World Dredging Congress, Amsterdam 1995.

Riddel J (1982). The development of the bucket ladder dredger on the River Clyde. Proc. Institution of Civil Engineers, London, Part 1, 72, Nov., 633–652.

Seaman KJ (2003). UK perspective on resources and use of marine sand and gravel. Proc. EMSAGG Conference "European marine sand and gravel – shaping the future". February 2003, Delft, Netherlands.

Vellinga P (2001). Dredging's interaction with the environment. Paper presented at CEDA Dredging Days, Amsterdam, 2001. *Cited in* Burt TN (2002) So what do they think of us: Dredging seen from the outside looking in. Terra et Aqua No 88, Sept 2002.

Chapter 5

Dredging in Spanish Coastal Waters

Roberto Vidal
Contracts Director of DRAVOSA, Madrid, Spain

5.1 THE SPANISH COAST

From time immemorial, the Iberian Peninsula has been an important enclave in the history of civilization.

Surrounded by three seas, the Atlantic, the Mediterranean and the Bay of Biscay, the Iberian Peninsula has always been of vital importance for communication routes because of its eccentric location on the European continent. Important cities and ports such as Cádiz were founded by the Phoenicians in 1100 BC, Ampurias by the Greeks in 575 BC and Cartagena by the Carthaginians (227 BC). These ports connected the *Mare Nostrum* (the Mediterranean Sea) and the Atlantic Ocean. Later, with the discovery of America, the Spanish and Portuguese ports on the Iberian Peninsula continued to play an important role in history as the points of arrival and departure between the New and Old Worlds. Seville was the most important port during that time.

The Spanish coast is one of the most varied on the European continent. Roughly, the Iberian Peninsula is formed of contour lines that are almost straight with two large curves forming the Gulf of Valencia on the east and the Gulf of Cadiz on the south. The continental shelves, capes and bays mostly are not very pronounced.

The northern coast is noteworthy for its high, rocky terrain. Galicia's shores on the northwest of the peninsula are indented by estuaries (Ferrol, Coruña, Pontevedra, Vigo, etc.) where the sea deeply penetrates into the land. It is a unique example of a so called Ria-coast.

The coast southwest of the Straights of Gibraltar is formed of low, sandy land with long beaches and large rivers like the Guadalquivir, Guadiana and Odiel flowing seaward. The shores bathed by the Mediterranean Sea to the east of the Straights of Gibraltar are rough and rugged with rare exceptions such as the beaches on the *Costa del Sol* (Málaga).

The eastern coasts of Spain are full of variety. Stretches with cliffs such as the *Costa Brava* and areas of Denia and Villajoyosa contrast with delta areas (the mouth of the Ebro River), lakes of marine origin (Valencia's Albufera and the Mar Menor) and strips with dunes and long sandy beaches (southern part of Catalonia and provinces of Castellón, Valencia and Alicante). These contrasts are also found in the Spanish islands. The Balearic Islands (Mallorca, Menorca, Ibiza) has small coves with sandy beaches separated by high cliffs.

The seven main islands forming the Canary Islands are of volcanic origin: impressive cliffs formed of volcanic material contrast with long sandy beaches.

126 DREDGING IN COASTAL WATERS

Spain's marine activities take place within this geographic configuration that limits its structures, ports and coasts and, therefore, defines all the activities carried out there, including dredging.

Today, with 5,970 kilometers of extensive peninsular and insular shoreline, Spain's coasts continue to be of vital importance, which is a result of the great density of its population. Thirty-five percent of the country's inhabitants have their permanent residences in a 5 kilometer wide strip running along the shore (representing 7% of Spain's territory). The percentage increases to 80% during the summer months because of tourism. This fact, plus Spain's location on the edge of large areas of production and consumption, the importance of fishing and the historical insufficiency of surface transportation infrastructures on land have favored the construction of many ports.

Spain currently has 53 ports of general importance, managed by 27 Port Authorities, plus a significant number of smaller regional ports. (Figure 5.1). They handle 85% of the merchandise imported and 65% of the exports. The ports of general importance moved 377.9 million tons of cargo in 2003; the port of Algeciras in southern Spain being at the head with 58 million tons, followed by Valencia and Barcelona on the east coast with 34 million tons each.

Spain has more than 1,800 kilometers of attractive beaches with a wide variety of relief and landscape. The natural riches and the mild weather give them significant ecological value by providing the characteristics required for important ecosystems of specific flora and fauna. The quality of Spain's beaches has also had significant economic and social

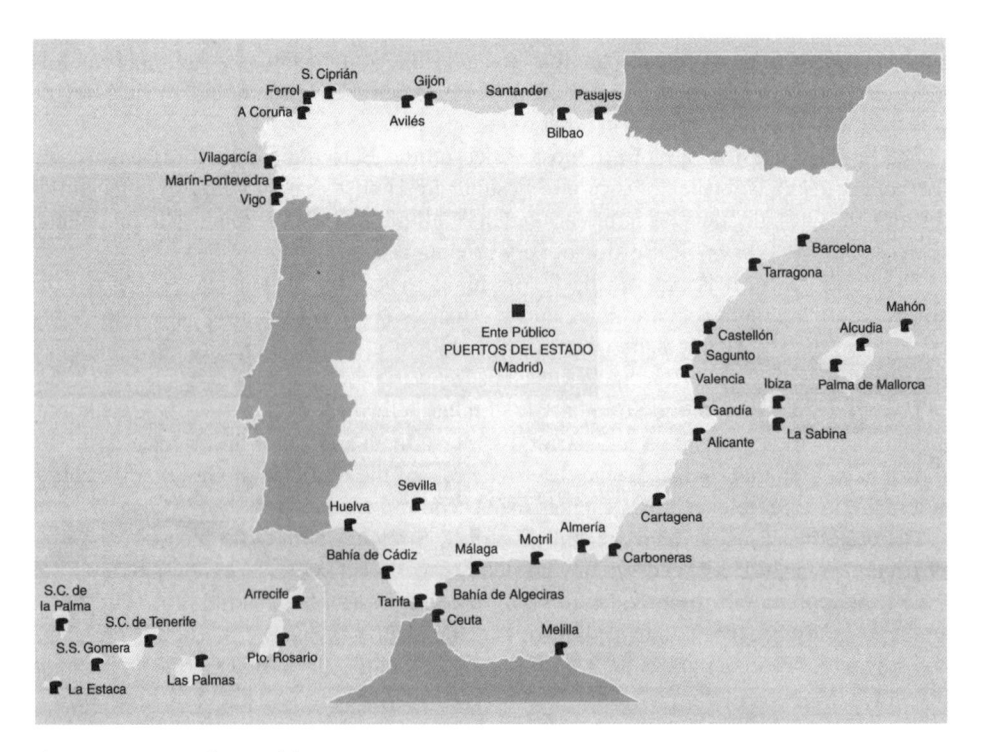

Figure 5.1. Ports of general importance.

consequences due to its effect on tourism. Spain is the second country in the world by the number of tourists it receives. Of them, 80% choose the coast as the destination for their stay in the country. The wealth that tourism contributes to the Spanish coastal areas creates more than one million jobs. For this reason, this sector is a basic pillar of Spain's economy.

Unfortunately, due to ignorance and neglegence, the Spanish shores began to suffer significant degradation decades ago, caused by buildings located too close to the coastline, artificial marine shelters and river dams. In the specific case of the beaches, the degenerative processes have been so irreversible that some are in danger of disappearing and huge quantities of sand have been required to maintain them artificially.

5.2 DREDGING IN SPAIN

5.2.1 Port dredging

The dredging activity carried out in Spain is mainly centered on port dredging.

Spanish ports can be divided into three groups depending on their geographic location:

1. River ports, the only one of significance being Seville located 90 kilometers upstream from the mouth of the Guadalquivir River which is the only navigable river of importance in Spain.
2. Naturally protected inland ports such as Huelva, Avilés, Santander and the ports in the estuaries of Galicia (Vigo, Marín, Ferrol, etc.).
3. Exterior ports, which are the large majority. For these, the lack of natural conditions have made it necessary to build breakwaters in order to reach the depths required and build platforms.

This classification makes it possible to explain the types of dredging that each port requires. The inland ports of Seville, Huelva, Avilés and Santander need constant maintenance dredging, while the dredging in the rest of the ports is capital dredging or, is as part of a project to build a new additional infrastructure such as a wharf, breakwater, etc.

Over the last ten decades there has been a constant, regular demand to dredge between 6 and 8 million m^3 per year. This regular annual average was greater in years when there was a large capital dredging project. Examples are the exceptional dredging jobs carried out in the port of Cádiz (8.2 million m^3) in 1991, in the ports of Huelva (6.5 million m^3) and Valencia (8.5 million m^3) in 1995 and in the ports of Valencia, Barcelona and Castellón (9 million m^3 total) in 2003. From 1975 up to and including 2003, 203.6 million m^3 were dredged in the 53 Spanish ports of general importance of which 135 million were capital dredging (66% of the total), and the rest of the volume, 68.6 million m^3, were for maintenance (34%). The smaller regional ports, that do not fall under the responsibility of the Port Authorities, generate a total volume of approximately 600,000 m^3 of dredging a year, mainly for maintenance purposes. In this 29 year period, the large Spanish ports that were dredged were Huelva (25 million m^3), Santander (23 million m^3), Valencia (29 million m^3), Seville (19 million m^3), Avilés, Bilbao and Tarragona (10 million m^3 each) and Cádiz (13 million m^3). Approximately half of the volume mentioned that was dredged during these 29 years

Maintenance dredging:
18.6 Mm³ 17%

Capital dredging:
92.3 Mm³ 83%

☐ Total dredged: 110.9 Mm³

Figure 5.2. Last decade dredging.

Dumped at sea:
26.4 Mm³ 24%

For reclamation or
beaches: 84.5 Mm³ 76%

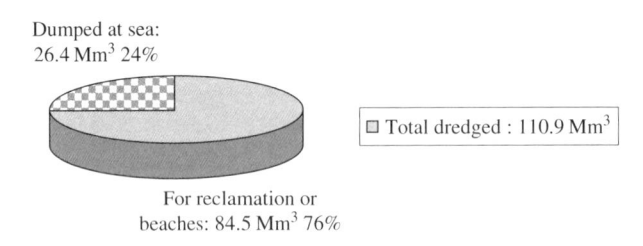

☐ Total dredged : 110.9 Mm³

Figure 5.3. Last decade dredging.

was done during the last 10 years: since 1994, 110,894,000 m³ have been dredged. Of this, 92.3 million was capital dredging and 18.6 million m³ for maintenance (Figure 5.2).

It must be pointed out that the volume of capital dredging during the last decade was 83% of the total. This percentage is much greater than the amount carried out since 1975 and is far more than dredging for maintenance. This is because of the need to increase the competitiveness of these ports and to adjust to the new, larger ships and their requirements for greater depths, more and better infrastructures and installations.

During the dredging work performed in the last decade, 84.5 million m³ of the sand has been used as infill for reclamation or for beach replenishment. This is 76% of the total volume dredged. The remaining 26.4 million m³ were dumped at high sea (24% of the total volume dredged). These figures reflect the tremendous infrastructure construction activity that has taken place during the past decade (Figure 5.3).

The projects that have involved the greatest volumes of dredging over the last 35 years are the following: Sevilla–Bonanza Canal Dredging, in Guadalquivir River, with 24 million m³ dredged by cutter suction dredger, (1969–1976); widering and deepening of the access channel and outer basin of the port of Huelva, 18 million m³ in several phases (1981–1983, 1995–1996 and 2001–2002); enlargement of the southern basin in the Port of Valencia (1994–1996) with 15.5 million m³; general dredging in the Port of Cadiz and beach regeneration (1991) involving 8.3 million m³ and the access and berthing for the tanker port in La Coruña (1972) with 5.5 million m³.

5.2.2 Beach nourishment

At present there is not very much dredging for other than Port works in Spain: besides a few underwater trenches for outfalls, the most relevant dredging jobs have been for beach

Figure 5.4. La Malagueta Beach (Málaga).

nourishment projects. The policy of renewing Spanish beaches with dredged sand was very intense during the eight years between 1988 and 1996 when more than 30 million m^3 of sand were dredged and placed on beaches. But since then regeneration with dredged sand has been very limited, due to restrictions on extraction permits.

During the period between 1988 and 1996, the most outstanding beach replenishment projects were the Maresme Beach (Barcelona, 1993–4) with 4.3 million m^3 of sand extracted and deposited, the beaches of Calafell, Vendrell and Roda de Bara (Tarragona, 1993) with 2.7 million m^3, the beaches of La Victoria (Cádiz, 1991) and La Malagueta (Málaga, 1990) (Figure 5.4) with 1.9 million m^3 each, and San Juan and Muchavista beaches (Alicante 1991) with 1.2 million m^3.

There have been few projects in recent years with the specific purpose of dredging sand from the sea bottom to replenish beaches. In the three year period 2001–2003, only 3.5 million m^3 were dredged for beach regeneration projects. This is very little when compared with the period between 1988 and 1996 when the yearly average was nearly 4 million m^3.

It must be pointed out, however, that in the port dredging projects mentioned above, part of the sand extracted went to replenish beaches. Thus, in the years 2001, 2002 and 2003, 1.3 million m^3 were used in this way. In any case, this activity represents a very small percentage of the total amount dredged in ports (4.2%) because of the poor quality of the material extracted.

5.3 SPANISH LEGISLATION ON DREDGING

Spanish legislation on dredging is not totally clear or complete and is subject to interpretation. This is partially because to a large extent it is a more or less literal transcription of the European Directives and they are not clear on this matter either.

Dredging is dealt with in several Spanish laws:

5.3.1 Coastal Law 22/1988

Article 63 of this law deals exclusively with dredging along the coast and does not consider dredging performed in port service zones. It establishes that the extraction of aggregates and dredging shall be authorized after having *evaluated* the effects of this work on both the place of extraction or dredging as well as on where it will be deposited. It is specifically prohibited to extract aggregates along the coast for construction purposes except to create and replenish beaches.

This last point makes it impossible to use aggregates extracted from the sea bottom along the coast as infill in ports, for example. Although this Law of Coasts has not been modified, a later Law (48/2003) has made it possible to use materials dredged along the coast as infill in ports. Law 48/2003 is discussed below.

The need to analyze the possible effects of dredging on the environment is established more clearly in the Spanish laws on ports and harbors.

5.3.2 Law of Ports and Merchant Navy, 27/1992

Two of its articles establish legal requirements on dredging projects.

Article 21 sets forth a series of requirements that must be met by dredging projects. This article was later modified by law 48/2003 as discussed below.

Article 62 establishes that all port dredging *projects* must include a study evaluating the effects on the coastal dynamics, marine biology and possible location of archaeological remains.

5.3.3 Law on the economic regime of and services provided by ports of general interest (Law 48/2003)

This law modifies article 21 of law 27/1992 by indicating in its second paragraph that port works can include dredging and infill jobs that by their nature, final disposal or protective isolation do not cause polluting processes that surpass the levels stipulated by the law on marine water quality.

Article 131 deals specifically with dredging and dumping jobs and projects.

As for **Marine Works**, it is established that the execution of all *dredging and reclamation projects in port zones* shall be authorized by the corresponding Port Authority. It also requires the authorization of the Marine Authority when the dredging or dumping affect the safety of navigation in the port zone.

All *dumping to be done outside the port zone* must be authorized by the corresponding Coastal Authority.

Whenever dumping is to be done, studies or analyses must be carried out to evaluate the effects on the sedimentology of the coast and the submarine biosphere as well as the polluting capacity of the material dumped. These studies shall be submitted to the Departments of the Environment and Fishing for their report.

All dredging works carried out outside of port zones to obtain material for reclamation within a port must also be authorized by the Coastal Authority.

This text permits the possibility of using offshore dredging material as fill in port works which was prohibited by the Coastal Law which was previous to the law being commented on here.

All dredging *projects*, nevertheless, must include a study of how the dredged material will be managed and of the location of the dump site.

Port projects shall include, if appropriate, a study of the possible location of archaeological remains. This study will be submitted to the corresponding authority on archaeological matters.

Projects outside port zones will include a study evaluating their effects on the coastal dynamics and marine biosphere. This study will be submitted to the appropriate Departments of Fishing and the Environment.

When a *project is submitted and a procedure* to evaluate the environmental impact is started, the studies required in such procedure shall be reported on by the Marine Authority and by the Departments of the Environment, Fishing and Archaeology.

5.3.4 Law 6/2001, on the assessment of the environmental impact

This law is a modification of a previous one referring to Directive 85/337/EEC. It only mentions dredging in rivers and streambeds or protected wetlands, and marine dredging of sand. Therefore, it is very limited.

Annex I of this law establishes that an environmental impact evaluation is required for river dredging with a volume of over 20,000 m^3/year carried out in the river bed or in protected wetlands, and for marine dredging, when more than 3 million m^3 are extracted per year. The rest of sand extraction jobs in the marine environment only require an environmental impact evaluation if it is considered necessary by the competent authority.

To comply with the environmental impact *evaluation*, the first part of which is established in Royal Decree 1302/1986, an environmental evaluation is being used as screening. This procedure involves:

1. Submitting a summary report on the project
2. Consultation of the people or organizations that are affected
3. Information on the result of the consultations concerning the project; the competent environmental agency to determine if it will be necessary to make a complete, detailed study.

Article 2 of Law 6/2001 establishes the data that must be included in the environmental impact study:

1. Description of the project and foreseeable requirements on the use of the soil and other natural resources. Estimate of types and quantities of wastes, materials to be dumped and emissions.
2. Main alternatives studied and justification of the reasons why this particular solution was chosen.
3. Evaluation of the effects that the project will have on the people, wildlife, water, historical and archaeological elements, etc.
4. Measures foreseen to reduce, eliminate or compensate any significant effects.
5. Program of environmental surveillance.

Figure 5.5. Dredging and reclamation at Dársena Sur (Valencia).

5.3.5 European directives

Like many other member states of the European Union, Spain is affected by the European Directives on dredging operations.

The legal complexity of performing dredging work in ports in Spain, as explained above, has increased with the Directives on wastes and dumping and, recently, with the EU Water Frame Directives. The demands made to reach a good ecological level and to the study of water chemistry, without a doubt, will make it more difficult to perform dredging work.

Directives 75/442/EU, 1999/31/EC and 2001/118/EC on wastes and their dumping are not clear with respect to certain dredging operations, e.g. whether dredged materials should be considered as waste, any interpretation is possible. It is even possible to reach the extreme of having to prove whether the materials to be dredged are dangerous waste.

200/60/EC Water Frame Directive, included in the Spanish legal body by Law 61/2003, adds materials in suspension to the list of main pollutants.

5.3.6 International conventions to protect the environment

Spain has signed the following Conventions:

- **London Convention** to prevent pollution of the sea due to the dumping of wastes and other matter (1972).
- **OSPAR (Oslo/Paris) Convention** (1992) to protect the marine environment of the northeast Atlantic.
- **MARPOL Convention** (1973) the international convention for the prevention of pollution from ships.
- **Barcelona Convention** for protection of the Mediterranean Sea against pollution (1976).

As a result, Spain is adjusting its laws and measures to these agreements.

5.3.7 Recommendations for the management of dredged material in Spanish Ports

Many countries, such as the USA, Holland, Great Britain, Belgium, Sweden and Germany, have developed more specific regulations on environmental control in the case of dredging projects. In Spain, the "Recommendations for managing dredged material" (RGMD) have been applied in Spanish Ports since 1992. This document was written by CEDEX (Centro de Estudios y Experimentación de Obras Públicas). Although Spain signed the Conventions of London, Oslo/Paris and Barcelona, there were no Spanish regulations establishing the general principles as set forth by these conventions for granting permits for dumping dredged material. When the Law of Ports of 1993 (27/1992) entered into effect, it accelerated the need for a national regulation on this matter.

Recommendations to that effect were approved in 1994 with the consensus of all the official organisms implicated in this matter. Although the RGMD is not a law, there is a commitment to apply it in the field of competence of all official authorities. It is one of the only three documents in a language other than English (the other two are in Dutch) included in the bibliography of the report "Management of Aquatic Disposal of Dredged Material (1986)" of the PIANC, the Permanent Environmental Committee. The RGMD is formed to two very different parts. The first establishes the criteria for categorizing the material dredged based on the concentration of pollutants, and places restrictions on each category with respect to dumping it into the sea.

The second part is a guide on the studies necessary to adequately manage dredged material and, in particular, on how to classify the material in order to apply the criteria of the first part.

5.4 THREE EXAMPLES OF ENVIRONMENTAL ACTIONS

Three dredging projects carried out in Spain are described below. The first involves the dredging works performed recently in the Port of Huelva, a model of good management of the use and final destination of the products extracted. The second concerns the project for the Valencia Port Authority in Sagunto to dispose polluted materials in a previously dredged pit and cover them with clean sand. (Contained aquatic disposal). Finally, the Barcelona Port Expansion Project is discussed, which includes corrective measures along the coast in order to reduce the environmental impact of the expansion project.

5.4.1 Management of dredged material in the Port of Huelva

The Port of Huelva is located between the Tinto and Odiel Rivers in southwest Spain. A 23 kilometer long navigation channel leading to this port through the years had to be adapted to the different types and sizes of ships that use the Huelva wharves. The Port of Huelva mainly handles solid and liquid bulk cargo. The need to work with larger and larger gas tankers, containing liquid natural gas made it necessary to carry out a series of projects to deepen and widen various zones of the port in recent years. Furthermore, the Tinto and Odiel Rivers, that converge at Huelva, carry a large amount of suspended matter causing significant siltation in the navigation channel and wharve zones. For these reasons, the port has required large dredging projects in recent years. Numerous copper iron pyrite mines have been exploited throughout history in the catchment areas of both rivers upstream from the port. Material has been leached and carried from their slag heaps, causing an effluent loaded with solids and rich in heavy metals. The latter are soluble in an acidic medium and settle when they enter into contact with seawater. Thus, the heavy metals pollute the mud that has to be dredged later in the port. For these reasons, the Huelva Port Authority has an environmental policy on the handling of the sediment. They are a model of good management of dredged products.

The different characteristics of the products to be dredged have made it necessary to study and define a place for disposing of them in accordance with these characteristics. Following the CEDEX "Recommendations for managing materials dredged in Spanish Ports", each dredging campaign must be preceded by a study to determine the characteristics of the material to be dredged in order to classify it and determine what is to be done with it. This has helped to optimize the end use of the material extracted: the solution proposed for treating the polluted dredged products was to deposit them in a land-based confined disposal site. To date, three such sites have been built and a fourth is currently under construction.

The first site, defined as a pilot zone, is located on the right bank of the entry channel and has a capacity of 375,000 m^3. It was of an experimental nature. The second one is 2.5 km long and 200 meters wide and has a design capacity of 2 million m^3. There is a smaller, impermeable deposition site for materials that have a higher degree of pollution. The third site has a theoretical capacity of 1.5 million m^3, is 1,000 meters long and 220 meters wide.

In 1995, 6,500,000 m^3 were dredged. Almost 5 million m^3 were clean, coarse sand that was dumped just offshore from beaches so that it could be recovered later and used to replenish them. The rest of the sand, that is 800,000 m^3, had a moderate concentration of pollutants that required only soft protective insulation; it was placed in the spoil deposition site located on shore indicated above. The remaining 700,000 m^3 were clean sand but with more than 25% fines and was dumped in an adequate site in the open sea (Figure 5.6).

The dredging works carried out later, in 2001 and 2002, were also examples of good management of dredged material. In 2001, the need to build new wharves adjacent to the second site mentioned above created the need to consolidate the site. It was done by preloading it with 600,000 m^3 of sand dredged that year. The rest of the material extracted,

Figure 5.6. Deposits for polluted products. Huelva Port.

approximately 300,000 m^3, was polluted and, therefore, deposited in an appropriate site specially prepared for it.

The dredging scheduled for 2002 is also noteworthy. Of the 3,277,000 m^3 dredged, 650,000 m^3 of excellent sand was used to nourish the Matalacañas and Ayamonte beaches located 10 and 30 miles from the entrance to the port, respectively. Furthermore, 900,000 m^3 of sand of adequate quality was used for beach regeneration (clean and with less than 25% fines) and was dumped offshore in a near shore zone of 10 to 12 meters water depth. This will be redredged and used to replenish beaches when the beach sand there has been washed away.

The rest of the material dredged (polluted material, approximately 1 million m^3) was dumped on shore. Therefore, between 1995 and 2002, a volume of approximately 11,000,000 m^3 was dredged for the projects mentioned above. Of this amount, 6,500,000 m^3 were used to nourish the coastal area, 600,000 m^3 for consolidating wharf surfaces by preloading, and over 2,000,000 m^3 of polluted material was deposited at special sites on the shore.

Most of these dredging jobs were carried out by trailing suction hopper dredgers that extract material from the bottom and transport it in their own hopper to the dump site. The material deposited in the sea was dumped by bottom opening hoppers while the sediment deposited onshore or on beaches was pumped through pipelines. Some of the other equipment used for these projects are the following dredgers: *Volvox Iberia* with a hopper capacity of 6,000 m^3, the *Volvox Atlanta* with 4,700 m^3, the *Volvox Scaldia* with 2,500 m^3 and the *Dravo Costa Blanca* with 1,450 m^3.

5.4.2 Specific dredging in the basin of the Port of Sagunto

The Valencia Port Authority is the Public Company responsible for managing and administrating the ports of Valencia, Sagunto and Gandía. Because of their geographic location on the eastern coast of Spain and in the center of the Mediterranean arch, these are the Mediterranean Sea ports that are closest to the Gibraltar-Suez axis.

The importance of these ports is demonstrated by the volume of traffic that they handle which surpassed 34 million tons in 2003, with over 1.9 million TEU's (Twenty Equity Units). The Port of Sagunto contributed 4 million tons to this total.

Aware of the importance of the environment, the Valencia Port Authority has developed the ECOPORT Project, a methodology for Environmental Management Systems in Ports aiming towards sustainable development. This awareness of the importance of the environment led the Valencia Port Authority to develop the project "Specific dredging to deepen the inner basin of the Port of Sagunto" in 2000. In previous dredging work, some of the material dredged led them to suspect that heavy metals and hydrocarbons might be found in approximately 100,000 m^3 of the material to be dredged for the project in question. It was highly probable that this pollution originally came from the steelworks that had been located in the area some 20 years before. The presence of this pollution made it necessary to approach the management of the dredged material in a different way while dredging and dumping it. With this in mind, samples were taken and tests were run to determine the physical and chemical characteristics of the materials to be dredged following the CEDEX "Recommendations for managing dredged material" mentioned above. The analyses determined that the materials pertained to a category that, according to said recommendations, could be disposed

Figure 5.7. "Razende Bol" at Sagunto Port.

of by controlled dumping at sea. After studying the different management alternatives, the Technical Staff of the Valencia Port Authority decided that the best solution would be to confine the polluted dredged material in a previously dredged pit in the port's inner basin and then cover it and seal it with clean material. The *Razende Bol*, an 1880 HP hydraulic backhoe type dredger was chosen as the most adequate unit for performing this special dredging work (Figure 5.7).

The hydraulic boom of the backhoe allows careful, controlled movement during the dredging operation. In addition to its great precision, the bucket hardly moves or vibrates when ripping the material from the seabed, so that there is a minimum dispersion of the material contained in the bucket when it is rising from the bottom to the surface and when it carefully empties its load into the hopper of the transport barge.

First, a receiving pit was dredged in the center of the port's inner basin to a depth of -11.5 meters. Its approximate area is 210×180 m and its depth 6 m. The 182,000 m^3 of material dredged to form the pit, were dumped in the sea in an area authorized for this purpose. Next, 126,000 m^3 of polluted material was extracted and deposited in the pit by split barges. A 1,000 meter long floating silt screen was used to control the material from spreading when being dumped into the pit. Furthermore, the dredging area of the backhoe bucket operations was surrounded by geotextil curtains supported by a continuous floating structure on the surface in order to prevent pollutants from spreading during the dredging operation. As a final operation, the polluted material in the pit was isolated by covering it with clean material. Approximately 45,000 m^3 of sand was dredged from the entrance channel and dumped over the polluted material until it was completely covered and sealed.

5.5 BARCELONA PORT EXPANSION PROJECT. CORRECTIVE MEASURES ALONG THE COASTLINE

For over 500 years, Barcelona has been located in one of the areas of greatest geographic and economic potential on the European continent. With a natural hinterland of more than 20 million inhabitants, the Port of Barcelona is today one of the most important ports on the

Figure 5.8. Barcelona Port Expansion.

Mediterranean Sea and is focused on becoming the southern gateway to Europe for cargo traffic. This goal has become today's Master Plan to enlarge the port so that by 2015 it will have twice the volume of general traffic and handle more than 3.5 million containers per year. The actions defined in the Declaration of the Environmental Impact of the Master Plan include the creation of a beach located to the south of the port expansion area by supplying 3.5 million m^3 of sand (Figure 5.8).

The purpose of this beach is to reduce the erosion that the expansion of the port of Barcelona will cause along the coastline to the south. The artificial beach will act as a source of sediments from this part of the coast, feeding the rest of the coast further south as it is eroded. Thus, it will act as a storage bank on which the forces of erosion can act, thereby preventing the coastline to the south from eroding, keeping it where it is today and increasing its stability.

It is foreseen that approximately 1,500,000 m^3 of sand will be provided with grain sizes of d50 > 200 micron covered by another 2,000,000 m^3 of d50 > 300 micron. The sand will be obtained from dredging in the port, from dredging in the beach area that will be occupied by the new part of the port, and from dredging the new riverbed of the Llobregat River to the south of the port expansion area. The coastline along which the sand will be placed, is located adjacent to wetlands of great environmental value and rich in wildlife that has been declared a Special Protected Area for Birdlife. Furthermore, it is an area with habitats of Community interest.To prevent the flora and fauna of the Protected Zone from being affected, the execution of the works will not be carried out at the times when the species are biologically most sensitive to them. Another requirement is that the construction methods to be used in the different operations and stages of sand replenishment, shall not affect the surface aquifer or the natural systems associated with it. For this purpose, a network of piezometers has been designed to control changes in the salinity and the level of the ground water of the aquifer, and a protocol of measures has been established in case deviations from the acceptable parameters are detected. Aspects related to the maximum acceptable levels of sound and light are also specified for the execution of the project.

REFERENCES

Autoridad portuaria de aviles (2000) – Proceedings of 1st "Seminario Internacional de Dragado" – Avilés – (Spain)
Autoridad portuaria de barcelona (1998) *"Actualización del Plan Director 1997–2011"* – Barcelona – (Spain)
Autoridad portuaria de valencia (2000) *"Memoria Anual"* – Valencia – (Spain)
Cedex (1994) *"Recomendaciones para la gestión del material dragado en los puertos españoles"* – Madrid – (Spain)
Fundacion dragados (2003) *"Gestión del Medio Ambiente"* – Madrid – (Spain)
Garcia navarro, pedro (2004) *"Gestión del material dragado en el Puerto de Huelva"* – Proceedings of 1st "Congreso Nacional de PIANC – Huelva – (Spain)
Ministerio de medio ambiente (1998) *"La costa de todos"* – Madrid – (Spain)
Ministerio de obras publicas y urbanismo (1993) *"Recuperación de la costa"* – Madrid – (Spain)
Puertos del estado (2001–2002–2003) *"Anuario Estadístico"* – Madrid – (Spain)
Vigueras Gonzalez, M. (1996) *"Dragas y Dragados"* – Puertos del Estado – Madrid – (Spain)
Vidal, R. (2003) *"Tres proyectos singulares de Dragado"* – Proceedings of 2nd "Intercambio Tecnológico Portuario" – Salvador de Bahía – (Brasil)

Chapter 6

Dredging in the United States

Robert E. Randall
Director, Center for Dredging Studies Ocean Engineering Program, Civil Engineering Department, A&M University College Station, Texas, USA

6.1 INTRODUCTION

6.1.1 Brief history of Corps of Engineers and US dredging

The United States Congress formed US. Army Corps of Engineers (USACE) in 1779 and its primary responsibility was coastal protection. In 1824, the General Survey Act authorized the use of Army engineers to survey road and canal routes, and Congress also authorized the employment of Army engineers to improve navigation on the Ohio and Mississippi Rivers that began the Army's long involvement in civil works activities. In the following decade, the Corps' involvement in civil works mushroomed from 49 projects and 26 surveys in 1866 to 371 projects and 135 surveys in 1882. Key developments occurred on the Ohio River that the Corps had dredged to a depth of 2.7 m by 1929, and on the lower Mississippi, where growing pressures for navigation and flood control led Congress to establish the Mississippi River Commission in 1879.

In 1899, Congress gave the Corps of Engineers authority to regulate obstructions to navigation. The River and Harbors Act of 1899 required a USACE permit for any work or structure in navigable waters of the United States and also required permits for placement of dredged material in navigable waters. The 1936 Flood Control Act recognized flood control as a proper activity of the federal government and gave responsibility for most federal projects to the Corps of Engineers. The Corps of Engineers took over responsibility for all Army construction in December 1941. This effort included military and industrial projects that involved more than 27,000 projects. After World War II, multi-purpose projects involving navigation, water storage, irrigation, power and recreation, in addition to flood control, predominated. The Corps' role in protecting the natural environment also expanded. The Fish and Wildlife Coordination Act of 1958 requires the USACE to consult with Federal and State fish and wildlife agencies to prevent damages to wildlife and provide for the development and improvement of wildlife resources for any proposed Federal dredging project in a navigable waterway.

The 1972 Clean Water Act requires the Environmental Protection Agency (EPA) in conjunction with the USACE to publish guidelines for the discharge of dredged or fill material such that unacceptable adverse environmental impacts do not occur. The responsibility for authorizing all discharges is assigned to the USACE and EPA guidelines must be applied. The National Environmental Policy Act (NEPA) requires that dredged material disposal activities comply with the NEPA requirements regarding identification and evaluation of alternatives. The Marine Protection, Research, and Sanctuaries Act of 1972 assigned the USACE the responsibility for authorizing the ocean disposal of dredged material using the criteria developed by the EPA. The EPA designates ocean disposal sites, but the USACE is authorized to select ocean disposal sites when an EPA site is not feasible or a site has not been designated. In 1986 the Water Resources Development Act (WRDA) created a financing arrangement for dredging associated with navigation maintenance and improvement projects. Local sponsors must provide one half the cost of the improvement and one half of the cost for additional maintenance dredging, and the Federal Government provides the other half. The Endangered Species Act of 1988 protects threatened or endangered species of animal and plant life.

The USACE Regulatory Program mission is to protect aquatic resources, while allowing reasonable development through permit decisions. The Corps evaluates permit applications for essentially all construction activities that occur in US waters, including wetlands. Permits are necessary for any dredging in US navigable waters. Views of other Federal, state and local agencies, interest groups, and the general public are considered during the permit process. Adverse impacts to the aquatic environment may be offset by mitigation requirements that may include restoring, enhancing, creating and preserving aquatic environment. Permit decisions should be made in a timely manner that minimizes impacts on the public.

6.2 US ENVIRONMENTAL LAWS AND ACTS

6.2.1 Environmental laws

Regulation of the placement of marine structures, dredging of ship channels, and the placement of dredged material within waters of the United States and ocean waters is a shared responsibility between the Environmental Protection Agency (EPA) and the US Army Corps of Engineers (USACE). The Marine Protection, Research, and Sanctuaries Act (MPRSA), the Clean Water Act (CWA), and the National Environmental Policy Act (NEPA) are the major US statutes/laws governing dredging projects, but a number of other US Federal Laws and Executive Orders must also be considered. Jurisdiction of MPRSA and CWA are illustrated in Figure 6.1. Procedures for evaluating dredged material that is proposed for disposal in ocean waters are governed by a 1991 joint USEPA and USACE document (1) and for disposal in inland or near coastal waters is governed by a joint 1996 document (2).

6.2.2 Brief summary of three major US laws or acts effecting dredging

6.2.2.1 *MPRSA – Marine Protection, Research, and Sanctuaries Act – Ocean Dumping Act-London Dumping Convention.*

The MPRSA is also referred to as the London Dumping Convention or the Ocean Dumping Act. Section 102 requires the EPA to develop environmental criteria in consultation with the USACE. Section 103 assigns the USACE responsibility for authorizing the ocean disposal of dredged material. The USACE must apply the criteria developed by the EPA. Section 102 gives authority to the EPA to designate ocean disposal sites. Section 103 authorizes the USACE to select ocean disposal sites for project specific use when an EPA site is not feasible or a site has not been designated.

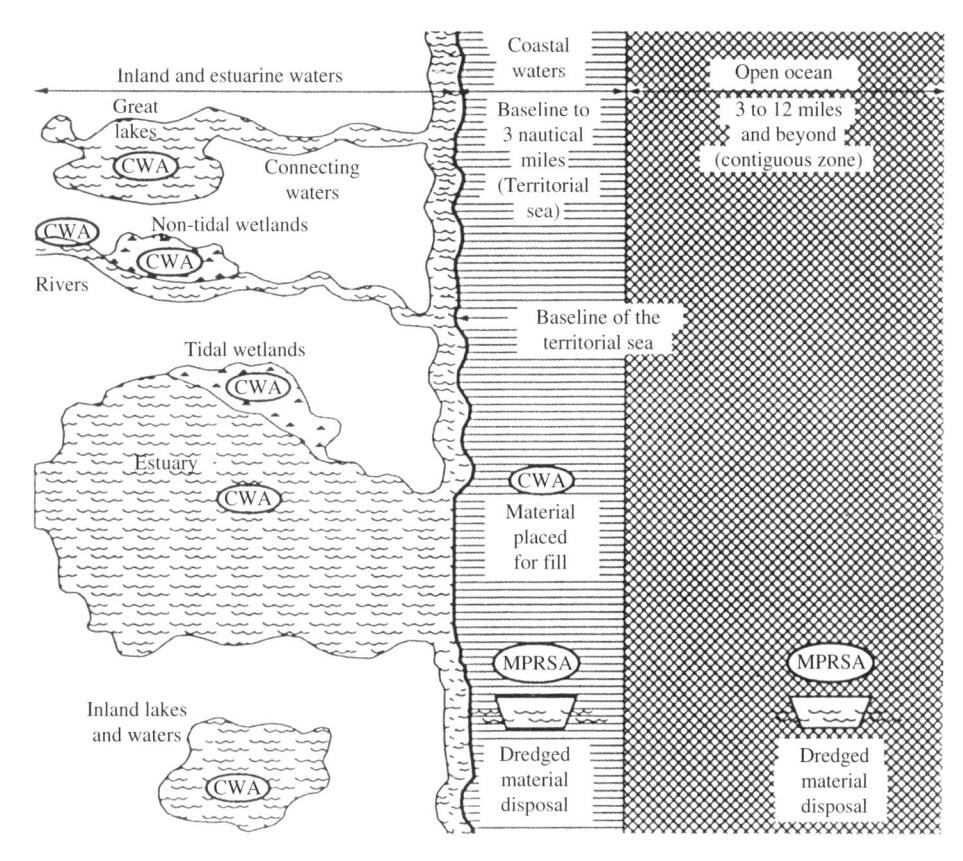

Figure 6.1. Jurisdiction boundaries for MPRSA and CWA (USEPA/USACE 1992).

6.2.2.2 *CWA – Clean Water Act*
The Clean Water Act is also called the Federal Water Pollution Control Act and Amendments of 1972. Section 404 requires the EPA, in conjunction with the USACE, to publish guidelines for the discharge of dredged, or fill, material such that unacceptable adverse environmental impacts do not occur. Section 404 assigns responsibility to the USACE for authorizing all discharges and requires the application of EPA guidelines. Section 401 provides the States a certification role for project compliance with the applicable State water quality standards.

6.2.2.3 *NEPA – National Environmental Policy Act*
Dredged material disposal activities must comply with the applicable NEPA requirements regarding identification and evaluation of alternatives. Section 102(2) requires examination of alternatives to the action proposed, and these alternatives are analyzed in an Environmental Assessment (EA) or Environmental Impact Statement (EIS). For USACE dredging projects, USACE is responsible for developing alternatives for the discharge of dredged material including all facets of the dredging and discharge operation, including cost, technical feasibility, and overall environmental protection. Compliance with environmental criteria of the MPRSA and/or the CWA guidelines is the controlling factor used by the

USACE to determine the environmental acceptability of disposal alternatives. The NEPA process is finalized in one of two ways. First, a Finding of No Significant Impact (FONSI) is the final decision document when an EA finds that the preparation of an EIS is not required. Secondly, an EIS is prepared, and the decision document is called a Record of Decision (ROD) that specifies the recommended action and discusses the alternatives considered (3).

6.2.2.4 *Other Environmental Laws, Acts, and Executive Orders*
Many other environmental laws and US Congress Acts affect dredging and are listed below:

- Coastal Zone Management Act
- Comprehensive Environmental Response, Compensation, and Liability Act (CERCLA) Superfund
- Endangered Species Act of 1988
- Federal Insecticide, Fungicide, and Rodenticide Act (1972)
- Fish and Wildlife Coordination Act of 1958
- Hazardous Material Transportation Act (1990)
- Marine Mammal Protection Act (1972)
- MBTA – Migratory Bird Treaty Act
- National Fishing Enhancement Act of 1984
- National Historic Preservation Act of 1966
- OPA-90 – Oil Pollution Control Act (1990)
- OSHA – Occupational Safety and Health Act
- Resource Conservation and Recovery Act (RCRA)
- River and Harbors Act of 1899
- Rivers and Harbors Improvement Act of 1978
- Rivers and Harbors, Flood Control Acts of 1970
- SDWA – Safe Drinking Water Act (1974)
- Submerged Lands Act of 1953
- TSCA – Toxic Substances Control Act (1976)
- Water Resources Development Act of 1986.

In addition, there are numerous Executive Orders that affect dredging projects such as Executive Order No. 11988 Flood Plains that requires consideration of alternatives to incompatible development in flood plains. Another order is Executive Order No. 11990-Wetlands that provides for protection of federally regulated wetlands. There are many other Executive Orders that may affect dredging applications and these are found in the Code of Federal Regulations (CFR).

6.2.3 Permitting

The permitting system of the Corps of Engineers that is used in the United States is briefly described. Three types of permits exist: individual, nationwide, and general. An individual permit is required for locating a structure, excavating or discharging dredged material in waters of the United States. Nationwide permits are issued for some smaller or minor water bodies, and general permits are issued for certain regions that may require specific notification and reporting procedures. The typical review process is illustrated in Figure 6.2, and the permit application process including an example is summarized in Herbich (2000).

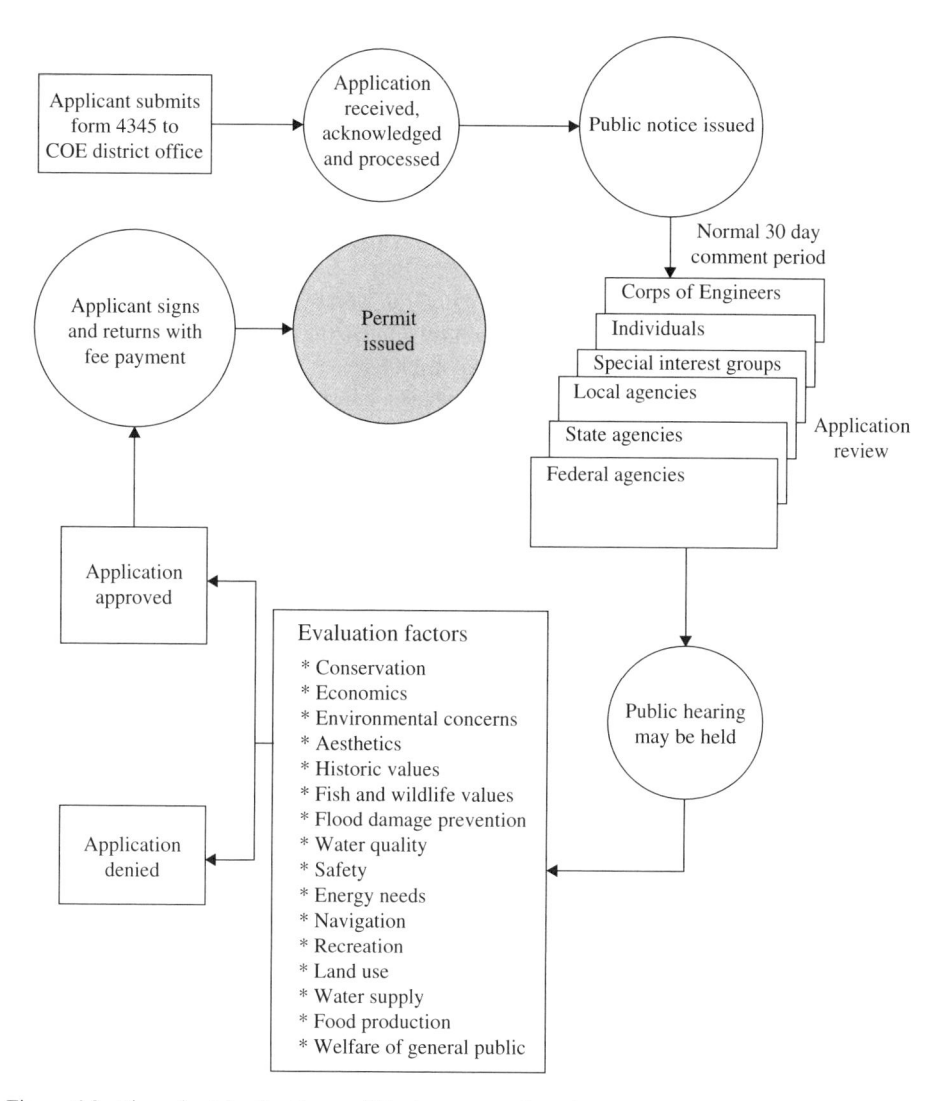

Figure 6.2. Flow chart for the Corps of Engineers permit review process.

6.2.4 Environmental impact statements

An environmental impact statement is often required to assess the impact of the implementation of a new engineering system or change to an existing system. The cost of these studies is borne by the person, organization, or agency requesting the new or changed system. Herbich (4) discusses the preparation of an Environmental Impact Statement.

6.2.5 Environmental windows

The National Environmental Policy Act in 1969 started the requests for environmental windows in the United States. These environmental windows are periods set aside for no dredging in order to protect sensitive biological resources or their habitats from the effects of dredging. Reine, Dickerson, and Clarke (5) conducted an extensive survey of districts and divisions of the US Army Corps of Engineers on environmental windows. Agencies in the US that most frequently request environmental windows include the Fish and Wildlife, National Marine Fisheries, Environmental Protection Agency, State Departments of Natural Resources, and State Fish and Game. Environmental windows are frequently imposed on dredging projects in both coastal and inland waterways. The physical disturbance to habitat and nesting is the justification for over 75% of the environmental windows in the US. Other justifications include sedimentation and turbidity, entrainment, vessel strikes, suspended sediments, fish migration, dissolved oxygen reduction, and recreational activities (hunting and fishing). Marine resources that are frequently cited as justification for environmental windows are anadromous fishes (salmon and striped bass), colonial nesting water birds (pelicans and terns), and endangered species (whales and sea turtles). The use of environmental windows increases the cost of dredging and often restricts the period of dredging to winter months when weather conditions are most dangerous for dredging operations.

In 2001, the Marine Board of the National Research Council in the United States recommended a six-step process for setting, managing and monitoring environmental windows (6). The recommendation included identifying stakeholders, project sponsors, science and resource experts, and engineering experts. Step 1 is to agree on procedures and a timetable. Step 2 considers project details and identifies resources of concern. Step 3 involves the science and engineering teams identifying the availability and validity of data, recommending dredging technology, recommending and prioritizing environmental windows. In step 4 the stakeholders and project sponsors review the recommendations of science and engineering teams, consider the socioeconomic implications, and set the windows. Step 5 includes dredging within the established windows, monitoring of project, and synthesizing the findings. Step 6 involves reconvening the stakeholders and project sponsors to review the above steps and recommend improvements.

6.3 JONES ACT

Dredges working in the United States must comply with Section 27 of the 1920 Merchant Marine Act (known as the Jones Act), the Foreign Dredge Act of 1906, and the Shipping Act of 1916. Several web sites listed as anonymous authors (7, 8, 9, 10 and 11) and Sickles (12) were used to summarize the Jones Act and its effect on dredging in the United States. The US Congress has upheld these acts repeatedly, and the majority of US citizens and the US government continue to abide by and support the Jones Act laws. In 1790, the US realized the need to protect its merchant marine and passed a tonnage act favoring US vessels. Although costly, it was still possible for foreign ships to become involved in US trade. The Foreign Dredge Act of 1906 required that a US vessel must be seventy five per cent owned by US citizens, that seventy-five per cent of the voting power must reside with US citizens, and that the chief executive officer and chairman must be US citizens in order for a ship to legally operate in US waters. The Jones Act was introduced in 1920 as part of the Merchant Marine Act of 1920, and it was merely a reinstatement of these pre-established laws, confirming the government's belief that domestic commerce and the US defense should be regulated by US citizens. Since the Jones Act was passed, several amendments have been added for clarification purposes.

The Jones Act is a "cabotage" law. Cabotage laws are found in almost every country that has a Merchant Marine fleet, and every country has its own shipping standards and regulations. The US is one of the only countries to require that ships must be built and repaired in the US in order to qualify as Jones Act ships. The US laws are much less flexible in permitting waivers. There are very few countries that operate without any cabotage laws governing shipping. Canada and Mexico, America's neighboring countries, require that a barge (or tug) be owned and operated by a national crew. Canada levies a rather high twenty five per cent tax on foreign vessels, which limits the number of foreign operating vessels involved in its domestic commerce trade.

Jones Act vessels receive no government subsidies, price supports or tax exemptions. The foreign flag vessels (non Jones Act vessels) receive subsidies in order to compete with foreign vessels that receive subsidies from their own countries. US-flag operators charge more than their foreign counterparts because they are required to pay Federal, State, and local treasury taxes. In order to protect the domestic trade it is necessary to disallow foreign competitors unless they are subject to the same taxes to which US-flag vessels are subject. US ships receive no government subsidies. Additionally, many foreign ships are built with substantial government subsidies and use third world crews. Foreign flag operators enjoy tax benefits and incentives that are not available to current US flag operators. If the Jones Act was removed, competition between domestic and foreign ships would never be fair, and foreign companies would likely dominate the current domestic shipping industry. Ultimately, the result would be a dramatic reduction in US involvement in its domestic shipping trade. Another benefit of the Jones Act is the current level of safety required in the US domestic and international shipping industries. The US Coast Guard is trained to inspect ships with higher standards than the International Maritime Organization currently demands. Opponents of the Jones Act feel that the Jones fleet is outdated, that the US shipping industry has dramatically declined, and that the need to revise the Jones Act is long overdue. Although American ship industries must pay higher prices to build and operate their vessels, they would not be competitive if foreign vessels were allowed to compete. The closed environment has its own competition that performs at its own level, apart from the international arena. Repealing the Jones Act does not appear to be on the US government's agenda within the near future. It is thus important to understand American maritime shipping laws, specifically the Jones Act and the 1916 Shipping Act. Those involved in the dredging industry should become familiar with the 1906 Dredge Act as well.

6.4 US DREDGING VOLUMES AND COSTS

The Navigation Data Center of the Institute of Water Resources at the US Army Corps of Engineers collects data on dredging contracts for all US navigable waterways and is publicly accessible through the web site: *www.iwr.usace.army.mil/ndc*. The historical cost data between 1980 and 2002 are illustrated in Figure 6.3 and Figure 6.4. The data show the actual cost of dredging remained about constant at 400 million dollars (M$) from 1980 to 1990 and then it increased to about 500 M$ from 1990 to 1997. Beginning in 1998, the actual cost rose steadily to approximately 900 M$ in 2002. The fiscal year is from October 1 to September 30. The average cost per cubic meter followed a similar trend ranging from approximately $2/m^3$ in 1980 to $5/m^3$ in 2002.

146 DREDGING IN COASTAL WATERS

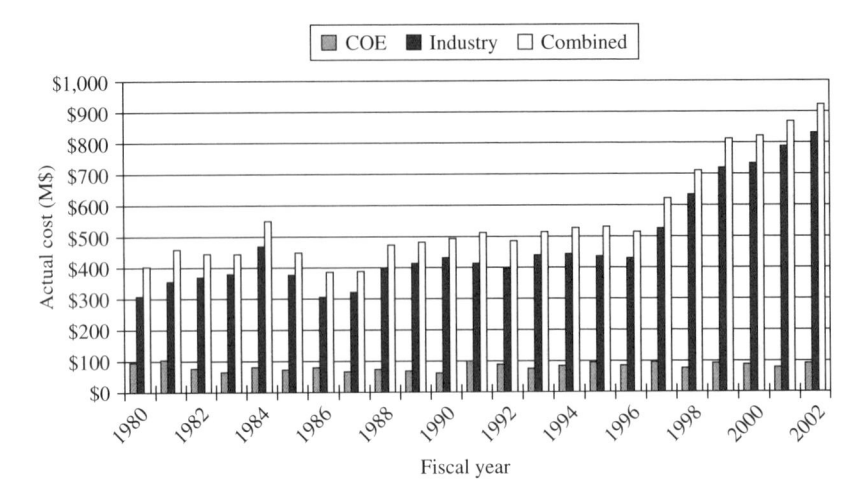

Figure 6.3. Actual dredging cost for the period 1980–2002.

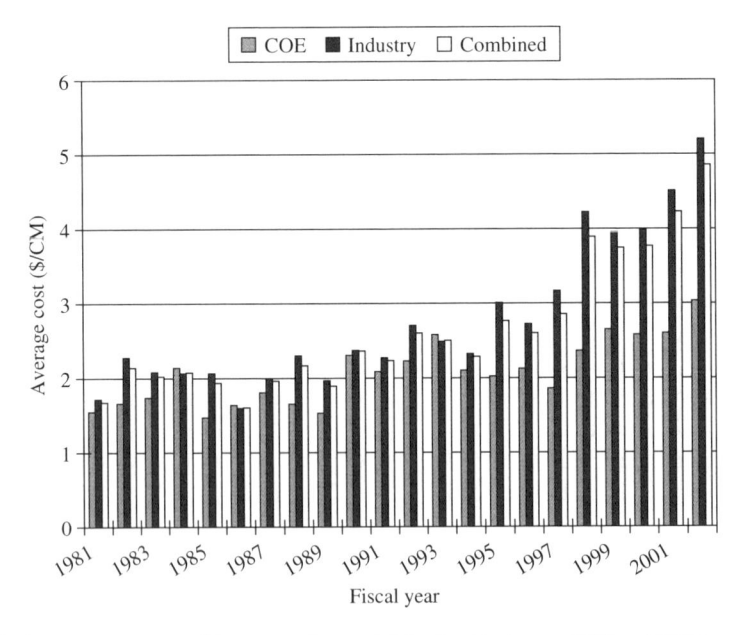

Figure 6.4. Average cost per cubic meter of material dredged between 1980 and 2002.

The amount of sediment excavated by dredging in the US over the period 1980–2002 is illustrated in Figure 6.5. Between 200 and 250 million cubic meters are dredged each year in the US. The Corps of Engineers (COE) operates a fleet of dredges, twelve dredges, that are primarily used for emergency situations or when industry dredges are not available. Corps of Engineers dredging amounts to approximately 20% of the annual dredging volume. The data indicate a small decrease in the dredging volume over the last 22 years. This

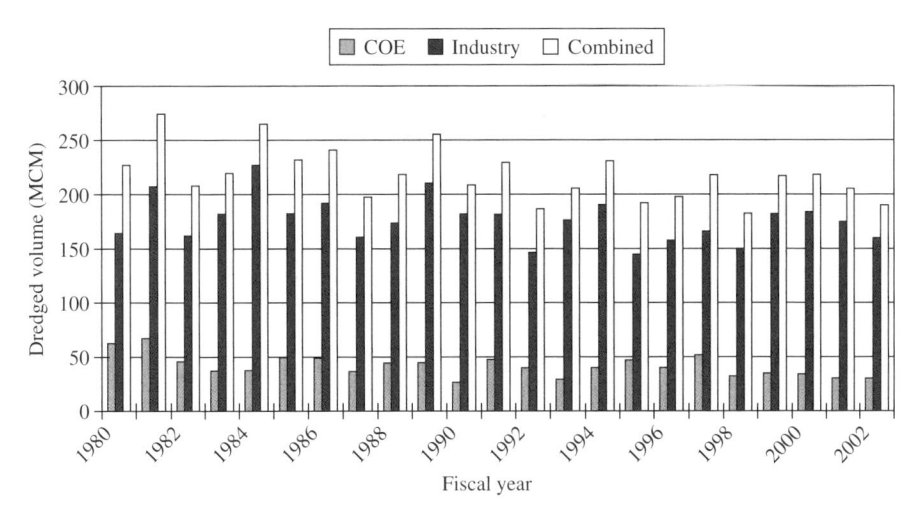

Figure 6.5. Total volume (cubic meters) of bottom sediments dredged in the US from 1980–2002.

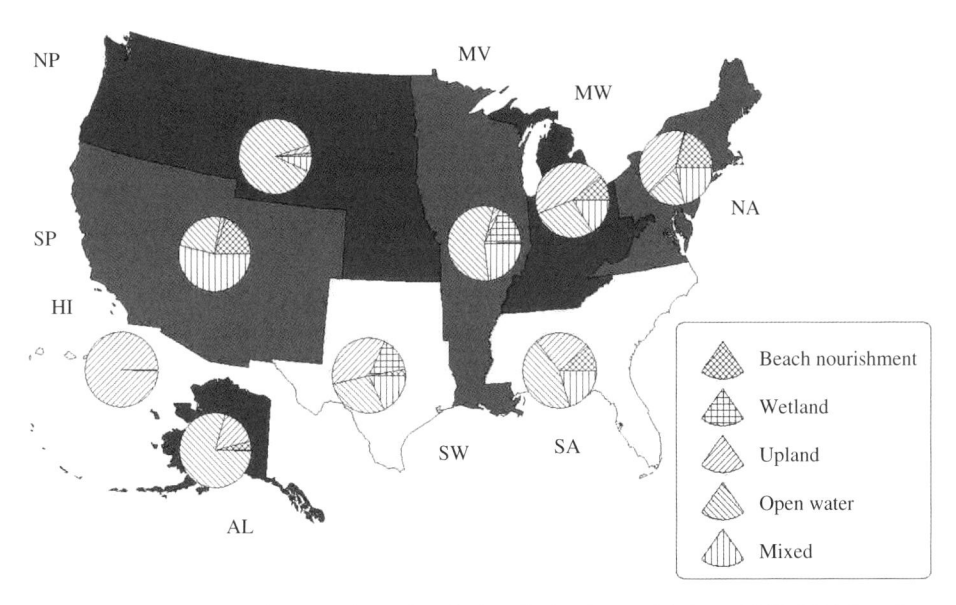

Figure 6.6. Distribution of the methods for placing dredged material from 1997–2001 based on volume (MCM).

decrease might be due to increased environmental restrictions such as dredging windows and shrinking federal budgets.

Over the five year period from 1997 to 2001, distribution by placement method is illustrated in Figure 6.6. The placement categories used by the Corps of Engineers includes open water placement, upland (confined) placement, and beneficial uses (i.e., beach nourishment and wetland creation). The placement methods are illustrated for different regions

148 DREDGING IN COASTAL WATERS

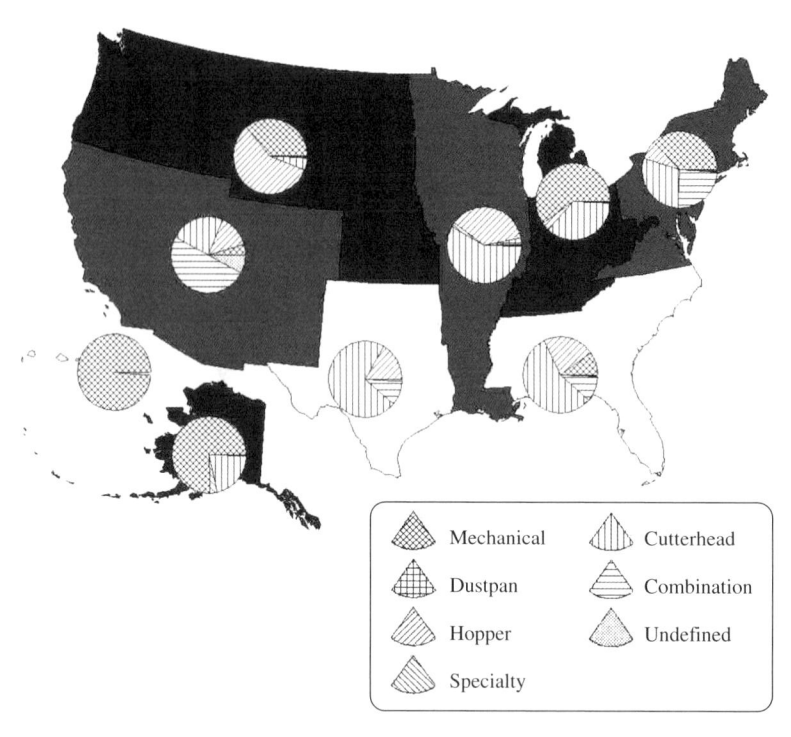

Figure 6.7. Type of dredge used for excavating dredged material for the period 1997–2001 based on volume (MCM).

of the United States that include Alaska (AL), Hawaiian Islands (HI), Northeast Atlantic (NA), Southeast Atlantic (SA), Mississippi Valley (MV), Midwest (MW), Southwest (SW), Southern Pacific (SP), and Northern Pacific (NP). Upland, open water or mixed placement are the most common placement techniques in all regions, and beach nourishment and wetland creation combined are generally small (e.g., less than 25% of the volume).

Figure 6.7 shows the distribution of the different types of dredging equipment used in different US regions. Mechanical dredging dominates the activity in Alaska and Hawaiian Islands. Hopper dredging is dominant in the Northern Pacific region. Cutterhead dredging is the dominant technique for the Southwest, Mississippi Valley, and Southeast Atlantic regions. The Northeast Atlantic region is pretty evenly divided between cutterhead, mechanical and a combination of techniques. Combination techniques dominate the dredging in the Southern Pacific region. Combination means at least two different dredging methods were used for a dredging contract.

The average annual dredging costs by region for the period 1997–2001 are illustrated in Figure 6.8 that shows the largest costs occur in the Southern Atlantic region followed by the cost in the Northeastern Atlantic with a cost of 174 M$/yr and 144 M$/yr, respectively. Figure 6.9 shows the distribution of dredged volume by region. The Mississippi Valley region (Mississippi River) shows the largest annual dredging volume of 54.3 million cubic meters (MCM). In the US, the Corps of Engineers classifies the type of dredging as new work, maintenance, both, and undefined. Maintenance dredging is shown (Figure 6.10) to dominate in all regions except for the Southern Pacific region.

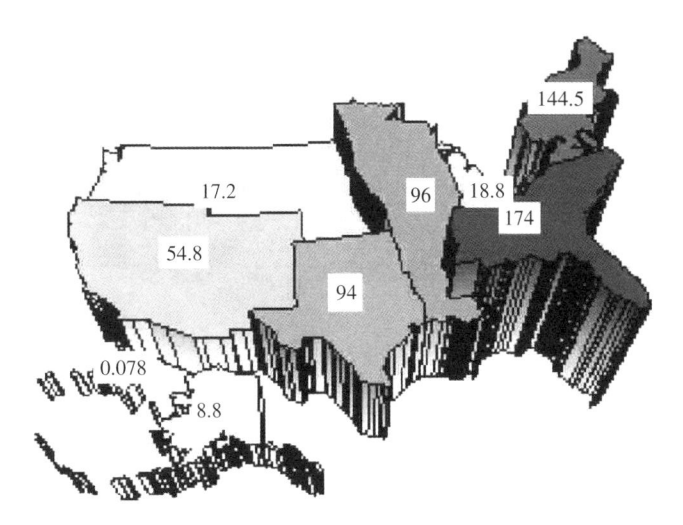

Figure 6.8. Average annual dredging costs over the period 1997–2002 in the United States (M$).

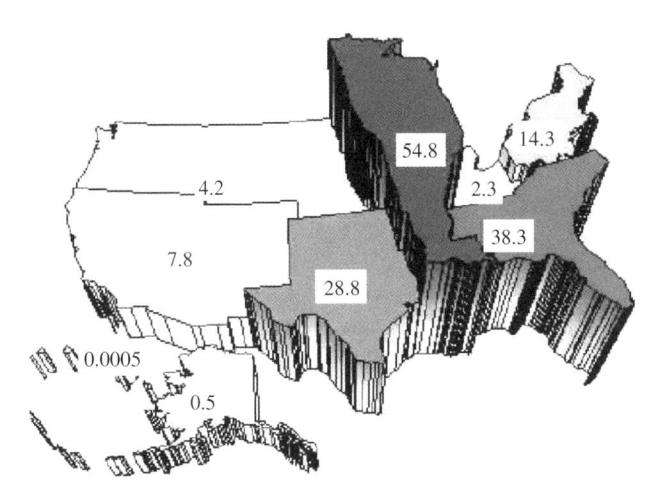

Figure 6.9. Annual volume of material dredged in the United States over the period 1997–2002 (MCM).

6.5 US DREDGING COMPANIES AND EQUIPMENT

6.5.1 Dredging companies

There are many dredging companies in the United States and every State has a least one dredging company. The World Dredging (13) and International Dredging Review (14) dredging directories are the primary sources of data for determining the distribution of dredging companies in the US as illustrated in Figure 6.11. A total of 288 dredging companies were identified, but it is expected that over 300 companies probably exist in the US alone. The largest US dredging companies include Great Lakes Dredge and Dock with

150　Dredging in Coastal Waters

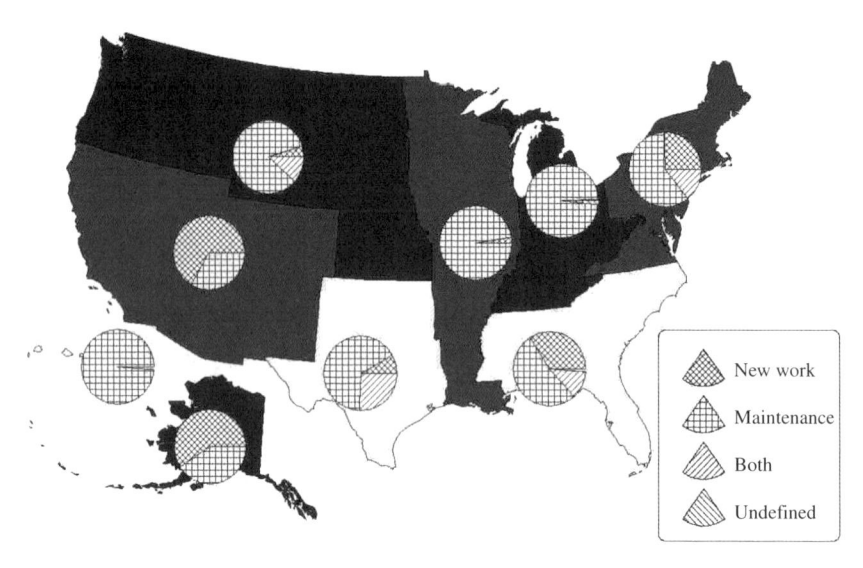

Figure 6.10. Volume of material dredged by work type from 1997 to 2001 (MCM).

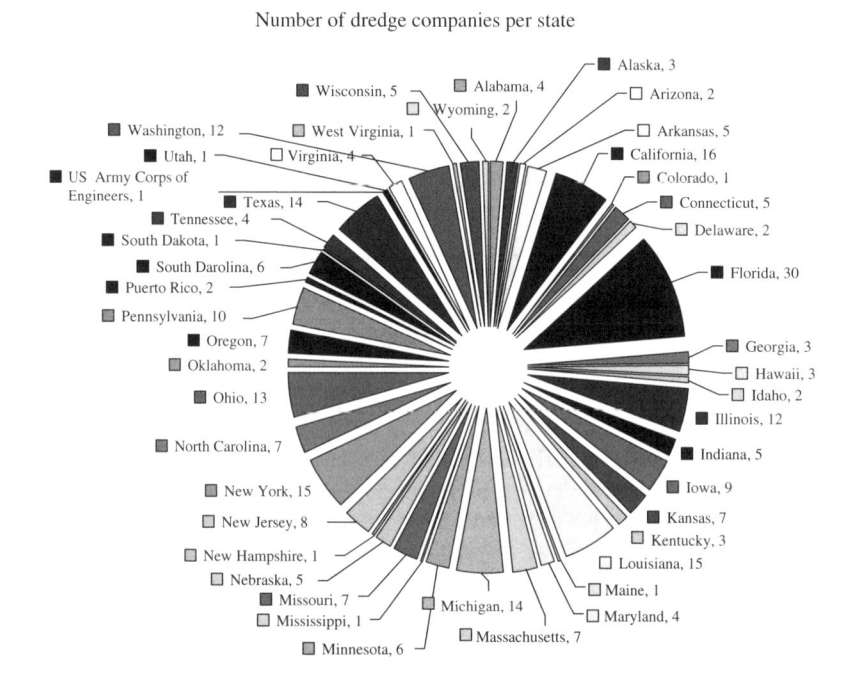

Figure 6.11. Distribution of dredging companies in the United States as of 2003.

corporate headquarters located in Oakbrook, Illinois which is just outside of Chicago. Other major US dredge companies include C.F. Bean Corporation, Weeks Marine, and Manson Construction that operate large cutter suction dredges and hopper dredges. The states with the largest number of dredge companies include Florida with 30, Louisiana and New York

US dredge distribution by state

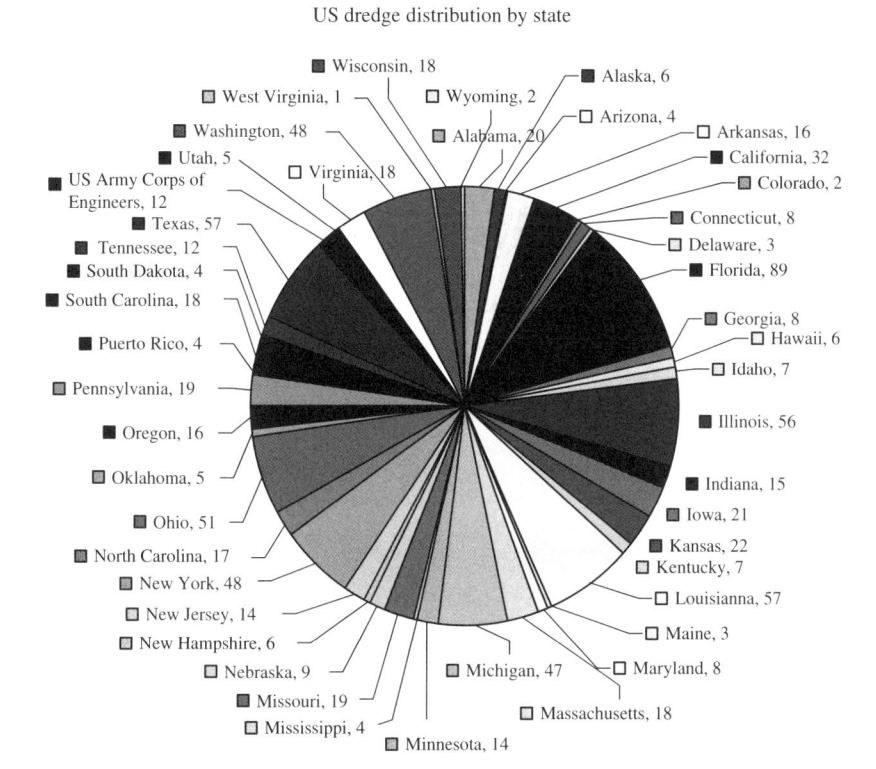

Figure 6.12. US dredge distribution by state in 2003.

with 15, California with 16, Texas and Michigan with 14, Ohio with 13, and Washington and Illinois with 12. The US Army Corps of Engineers operates a fleet of 12 dredges. Only the states of Rhode Island and North Dakota, Nevada, Montana, and New Mexico did not show a dredging company.

6.5.2 Types of dredges

A total of 905 dredges were identified in the United States, and the world population of dredges according to World Dredging (13) is 2460 dredges. Dredges may be classified as hydraulic or mechanical dredges. Hydraulic dredges use water to transport the dredged material (gravel, sand, silt and clay) through pipelines or in hoppers from the dredging site to the placement area. Common classifications of hydraulic dredges are cutter suction, auger, trailing suction hopper, plain suction, sidecasting, and water injection dredges. Mechanical dredges use a mechanical means of excavation and place the dredged material in barges, scows, or trucks for transportation to the placement area. Mechanical dredges do not use water to transport the dredged material as slurry. Common classifications of mechanical dredges include the bucket ladder dredge, clamshell, dipper, dragline bucket, backhoe bucket, and chain ladder dredges. The distribution of dredges by state is shown in

152 DREDGING IN COASTAL WATERS

US dredge distribution by class

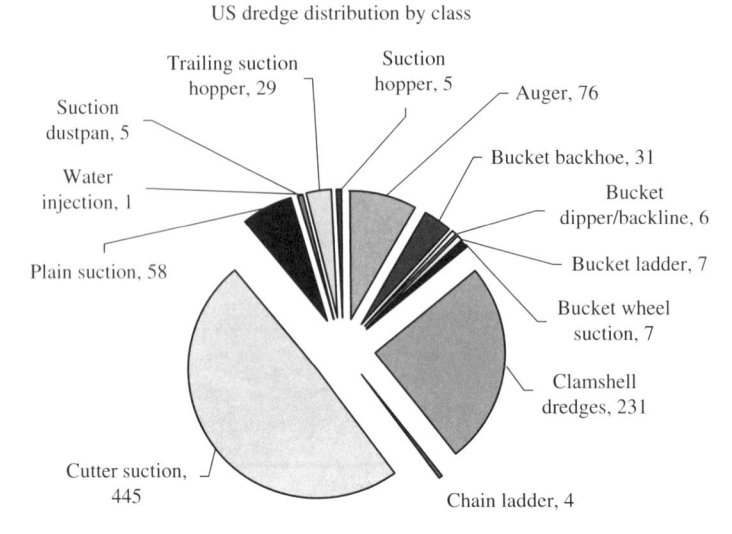

- Trailing suction hopper, 29
- Suction hopper, 5
- Suction dustpan, 5
- Water injection, 1
- Plain suction, 58
- Cutter suction, 445
- Auger, 76
- Bucket backhoe, 31
- Bucket dipper/backline, 6
- Bucket ladder, 7
- Bucket wheel suction, 7
- Clamshell dredges, 231
- Chain ladder, 4

Figure 6.13. US dredge distribution by classification in 2003.

Figure 6.12 that shows the largest number of dredges is found in Florida (89), Louisiana and Texas (57), Illinois (56), Ohio (51), and New York and Washington (48). No dredges were located in Rhode Island, Montana, North Dakota, Nevada, and Vermont.

Figure 6.13 shows the US dredge distribution by classification and the total number of dredges identified is 905. The total number of hydraulic dredges in the US is 619 that include 445 cutter suction dredges and 34 hopper dredges. The work horse of the US dredging industry is clearly the cutter suction dredge. The small dredges are dominated by the 76 auger dredges that are portable dredges that work in inland lakes and small waterways. The 286 mechanical dredges are dominated by the clamshell dredge with a total of 231 dredges in 2003.

Dredges are also classified by the size of the discharge diameter for the cutter suction dredges, volume of the hopper for hopper dredges, and the volume of the bucket (clamshell) for mechanical dredges. This distribution for US dredges is illustrated in Figure 6.14. Small cutter suction dredges are consider to have a discharge diameter of 305 mm or less and auger dredges are commonly included in this category. A majority of the cutter suction and auger dredges in the US are in this size classification. Large cutter suction dredges have a discharge diameter between 305 and 610 mm, and the very large dredges are greater than 610 mm. Of the 34 trailing suction hopper dredges in the US, most have a hopper volume less than or equal to 4,587 m^3 and only 8 hopper dredges are in the large classification with greater than 4,587 m^3. The largest hopper dredge is Great Lakes Dredge and Dock's Liberty Island that has a hopper volume of 7,646 m^3. For mechanical dredges, the size classification is based upon the bucket volume. Small mechanical dredges have a bucket volume less than or equal to 6.2 m^3 and this is the majority of the dredges. The large class has a bucket volume ranging between 6.2 and 12.4 m^3 that accounts for 50 dredges, and the very large mechanical dredges have a bucket volume exceeding 12.4 m^3. The largest US mechanical dredge is an 18 m^3 bucket.

Distribution of US dredges by size

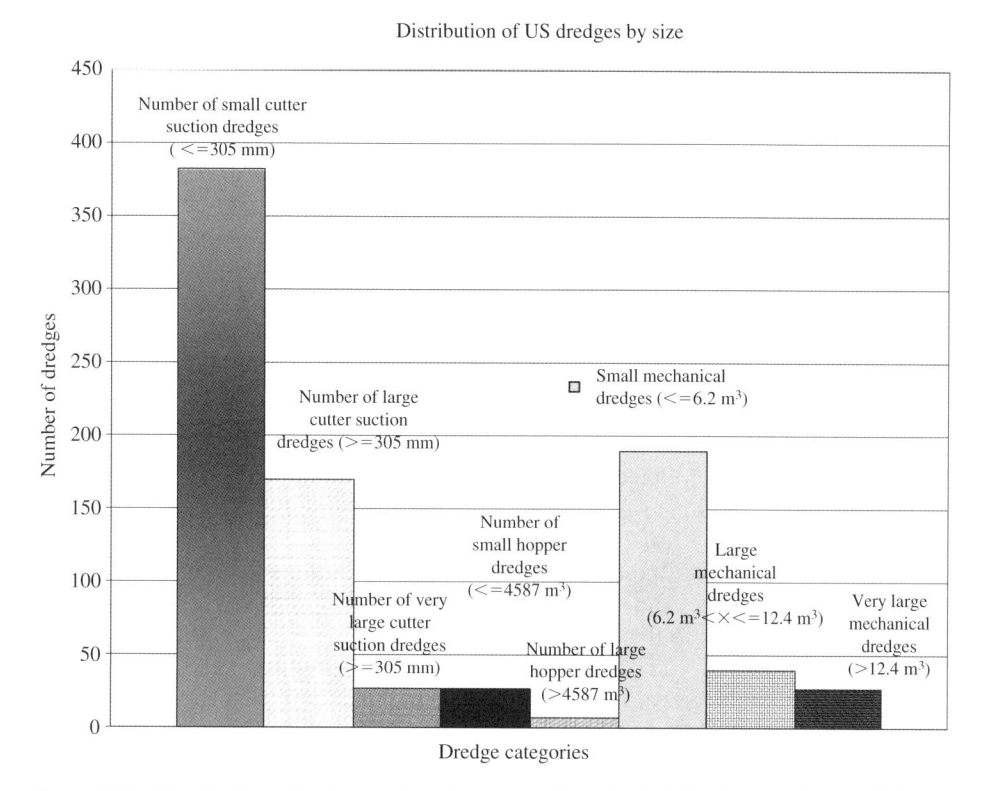

Figure 6.14. Distribution of cutter suction, hopper, and mechanical dredges by size of discharge diameter, hopper volume, and bucket volume, respectively.

6.6 CONTAMINATED SEDIMENTS AND CAPPING IN US PORTS AND WATERWAYS

Some US ports contain contaminants in the bottom sediments as a result of waste material from urban, industrial, river and navigation sources. Thus, dredged material from waterways, ports, and marinas must be tested to determine if the material is contaminated. In some cases the dredged material is contaminated and requires special handling during dredging and placement of the dredged material. Special dredging equipment has been developed for dredging contaminated sediments. Placement options have also been developed to manage the dredged material placement in open water disposal sites, confined disposal facilities, and for beneficial uses.

National and international laws, regulations, and treaties regulate dredging and dredged material placement. Individual countries have their own regulations and laws that apply to inland and coastal water. In the United States, these regulations or laws include the Comprehensive Environmental Response, Compensation, and Liability Act of 1980 (CERCLA), National Environmental Policy Act (NEPA), and the Clean Water Act of 1972 (CWA). Many countries have similar regulations related to dredging and placing contaminated dredged material in an environmentally sound manner. The Marine Protection, Research and

Sanctuaries Act (MPRSA), which is also called the London Convention, applies to the placement of waste in ocean waters outside the 22.2 km limit and except for fill it applies to waters outside the 5.6 km limit.

When contaminants are present in the bottom sediments benthic organisms are exposed and can ingest the contaminants and pass them onto the marine life that feeds on these benthic organisms. The bottom sediments may be resuspended by dredging, fish trawling, and natural mechanisms (e.g., hurricanes, cyclones, and other meteorological events) that expose marine life in the water column to the contaminants. As a result, marine life can ingest these contaminants that make them unsuitable for human consumption. Areas where contamination is probable include ports and coastal regions that are affected by urban and agricultural runoff, municipal and industrial waste streams, and other sources of pollution. The contaminants include heavy metals (e.g., lead, mercury, and cadmium), polynuclear aromatic hydrocarbons (PAHs), DDT, and polychlorinated biphenyls (PCBs).

The extent of contamination of water bodies varies around the world. In the United States, the Environmental Protection Agency estimates that approximately 10 per cent of the nation's water bodies are sufficiently contaminated to pose risks to fish and humans and wildlife that consume fish. It is estimated that 9.2 billion cubic meters of total surface sediments are where benthic organisms live. About 229 million cubic meters of sediments are dredged from the ports and ship channels annually and about 2.3 to 9.2 million cubic meters are contaminated to the extent that special dredging and disposal techniques are required.

Controls and technologies are required to handle contaminated sediments dredged from these water bodies. The Marine Board (15), which is part of the US National Academy of Engineers, conducted a study regarding the status of contaminated sediments. A comparison of the approaches used for the handling of contaminated sediments was evaluated related to feasibility, effectiveness, practicality, and cost was conducted. The approaches included in the study included interim control (administrative and technological), long term control (in-situ natural recovery, in-situ capping, in-situ treatment), sediment removal and transport, ex-situ treatment (physical, chemical, thermal, biological), and ex-situ containment. The study compared contaminated sediment remediation technologies that included natural recovery, in-place containment, in-place treatment, excavation and containment, and excavation and treatment. The report indicated the most commonly used technology is excavation and treatment at a cost of 20 to 100 US dollars per cubic yard. The most expensive technology is excavation and treatment at a cost of 50 to 1000 US dollars per cubic yard. In place containment is rapidly developing and its cost is estimated to be less than 20 US dollars per cubic yard.

6.6.1 Equipment

Special equipment for dredging contaminated sediments includes watertight clamshell buckets, pneumatic dredges, dredges using positive displacement pumps, conventional dredges with covers to reduce resuspension. Several dredging companies have watertight clamshell buckets that do not allow the water or sediment to leave the bucket until it is placed in the barge. The Cable Arm bucket is watertight and makes a level cut on the bottom. Pneumatic dredges have been developed primarily in Japan for soft sediments, and the amount of water that must be treated is greatly reduced. In most cases, these special dredges have much smaller production rates when compared to conventional equipment used for dredging clean sediments.

6.6.2 Placement options

Contaminated sediment may be placed in open water provided the sediment is capped with a layer of clean material. In some situations the contaminated sediment in a waterway is isolated from the water column by placing clean cap material over the contaminated sediments, and this is called in-situ capping. Dredging of contaminated sediment by hydraulic and mechanical dredges may also transport the contaminated sediments to a confined disposal facility (CDF) where the water and sediment must undergo treatment to decontaminate both the sediment and the water. Another possibility is to dewater the dredged contaminated sediment and use the contaminated sediment in a beneficial way such as in construction bricks, fill material that is covered with a layer of clean material, and concrete aggregate.

6.6.3 Capping

Guidelines (16) have been developed for capping contaminated sediments in open water. These guidelines contain case histories of capping projects and appendices the describe models used for modeling capping processes. In the United States the regulation that govern the placement of the contaminated dredged material and subsequent cap material are the Marine Protection, Research, and Sanctuaries Act (MPRSA). The Clean Water Act (CWA) of 1972 regulates dredged material and/or fill material placed in US waters (inland of the baseline to the territorial sea). In addition to the CWA and MPRSA regulations, the requirements of the National Environmental Policy Act (NEPA) must be satisfied, and there are also some US Federal laws, Executive Orders, and others that must be considered. Many other countries are signatories to the London Convention from which the MPRSA regulations were developed and therefore have similar requirements to meet regarding contaminated sediments and capping.

Capping may be defined as the controlled placement of contaminated material at an open water placement site, followed by a covering of clean isolating material called the cap (16). The capping project is an engineered project that requires a team approach that includes the input from engineers, biologists, chemists and dredge operators. The cap is design to physically isolate the contaminated material from the benthic environment, reduce the flux of dissolved contaminants to the overlying water column, and prevent resuspension and transport of the contaminated material. There are two types of capping for open water sites that are called level bottom capping (LBC) and contained aquatic disposal (CAD) and illustrated in Figure 6.15. The purpose of level bottom capping is to place a mound of contaminated sediment on a flat or gently sloping sea floor and then cover it with clean isolating material such as sand. For a contained aquatic disposal (CAD) site, the contaminated material is placed where there is lateral confinement such as a borrow pit, natural depression, or a subaqueous berm that confines the spread of the contaminated sediment, and a cap of clean material is placed over the contaminated sediment. The main advantage of the CAD site is the reduction in the amount of required capping material. Another form of capping is called in-situ capping (17) that means the contaminated sediments are located on the bottom of a waterway and there is a need to isolate the effects of the contaminants

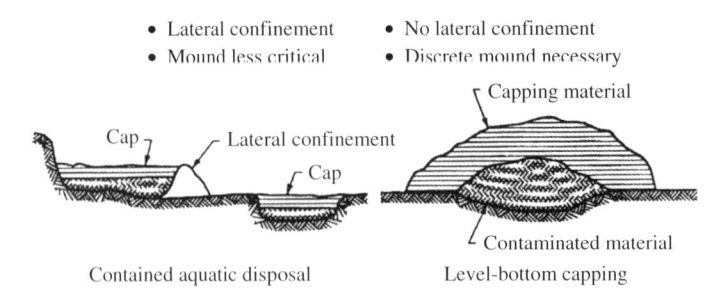

Figure 6.15. Schematic of capping contaminated sediment (16).

from the benthic organisms and the water column. In this case, clean material such as sand is placed over the contaminated sediments without dredging the contaminated sediments. The cap isolates the contaminants from the benthic biota and prevents the contaminants from entering the water column.

Generally, capped disposal sites are located in designated disposal sites that are nondispersive, meaning the sediments remain inside the disposal site boundaries. The selection of a capping site is guided by the same conditions as any other nondispersive open-water disposal site. However, the most desirable capping site is one that is considered a low-energy environment (e.g., low wave and current) where there is low potential for erosion of the cap material. The compatibility of the contaminated dredged material and the cap material is important for a successful capping project. When the contaminated material is dredged mechanically and placed with split hull barges, then the capping material can be placed by mechanical or hydraulic methods. However, when the contaminated material is fine-grained and placed by hydraulic methods, then only hydraulic placement of the capped material should be considered because of the low strength of the contaminated material. The exception is the slow controlled placement of a sand cap using controlled barge hull openings, submerged spreaders, and hopper pump-out with splitter plate.

The scheduling of the contaminated material placement and the subsequent cap placement must satisfy environmental and operational requirements. Once the contaminated material is placed on the sea bottom or in a contained aquatic disposal site, there is some necessary lag time preceding the placement of the cap. The important factors for determining the appropriate exposure time are the time required for benthic recolonization of the site, time required for self-weight consolidation, potential for effects prior to capping, and any monitoring requirements needed before the cap is placed. A period of 2 to 4 weeks is commonly allowed.

For level bottom capping, the dredging and placement of the contaminated sediment needs to result in a compact mound that is easily capped. The compact mounds usually occur when the dredged material is near its in-situ specific gravity. Mechanical dredging and point placement from barges are the most common technique for creating compact mounds. Contained aquatic disposal (CAD) projects control the lateral movement of the contaminated sediment when it is placed by either mechanical or hydraulic techniques, and the amount of cap material is minimized. When the contaminated material is placed hydraulically, then sufficient time before capping is needed to allow for some settling and consolidation to prevent mixing of the contaminated and cap material.

The design of the cap requires selecting cap material and determining the cap thickness. The cap physically isolates the contaminants from the benthic environment and the water column. For level bottom capping, the cap material consists of relatively large volumes of clean dredged material (e.g., sand) from other dredging projects or from a nearby source. Smaller volumes of clean sediments are needed for contained aquatic disposal sites and in-situ capping sites. The minimum thickness of the cap depends on the chemical and physical properties of the contaminated sediment and the cap sediment, water wave and current conditions, bioturbation of cap by benthic organisms, amount of consolidation, chemical isolation requirements, erosion, and other operational considerations. The total cap thickness is the sum of the bioturbation thickness, consolidation thickness, erosion thickness, operational thickness, and the chemical isolation thickness. In the recent history of capping projects the total cap thickness has been approximately 1 m of sand. When level bottom capping is used, the edge of the contaminated sediment mound often has only a thickness of 2–3 cm or less. Covering the edge-contaminated sediments with the total cap thickness is usually not done and a lesser thickness is placed. In some capping projects, intermediate caps may be desirable when multiple contaminated sediment placements are anticipated.

Level bottom capping, contained aquatic disposal, and in-situ capping sites require monitoring to insure the cap is constructed as designed and that the long-term integrity of the cap is maintained. Monitoring is necessary before, during, and after the placement of the contaminated sediment and cap material. Guidance for developing monitoring plans is found in (15) and (16). Science Applications International Corporation (SAIC) (18) describes a site specific monitoring plan for the New England Division of the Corps of Engineers that has been used to evaluate capping operations for over 20 capping projects. Multi-tiered monitoring programs are normally required for capping sites. Tiered monitoring programs have unacceptable environmental thresholds, sampling design, testable null hypotheses, and management options for conditions that exceed thresholds. Multidisciplinary advisory teams are normally employed to obtain the best technical advice in administering and completing the monitoring plan.

6.7 MAJOR US PORTS AND WATERWAYS

The major ports in the United States are located along the East, West, and Gulf coasts and in the interior of the US associated with the major rivers of the Mississippi, Ohio, Missouri and others. The top 15 ports based on tonnage is illustrated in Figure 6.16 with the ports accounting for the most tonnage being Port of South Louisiana and Houston, Texas along the Gulf of Mexico coast, New York on the East Coast, and the Ports of Long Beach and Los Angeles, California on the West Coast. Ports are also located in cities bordering the Great Lakes with access through the St. Lawrence Seaway. All of these ports require frequent dredging to keep channels at designated water depths for the deep draft vessels. The Port of South Louisiana stretches 87 km along the Mississippi River, and over 4,000 ocean going vessels and 50,000 barges use the port each year. This port is the top ranked port in the US for export tonnage and total tonnage, and it ranks third in the world. The Port of Houston (Figure 6.17) is ranked second in total tonnage and sixth in the world. A total of 6,414 vessels visited the port in 2002.

158　Dredging in Coastal Waters

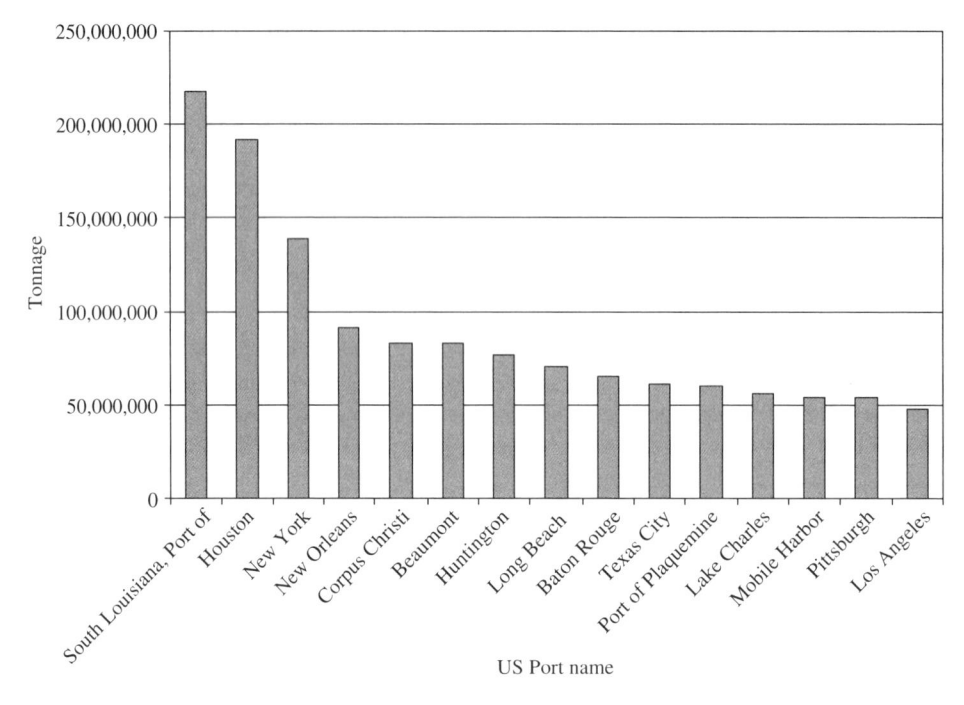

Figure 6.16. Top 15 US Ports in terms of tonnage.

Figure 6.17. Ports in the United States (19).

Figure 6.18. Rivers and Waterways of the Continental United States (GIWW – Gulf Intracoastal Waterway, AIWW – Atlantic Intracoastal Waterway).

The major rivers and waterways in the United States are illustrated in Figure 6.18 (19). The major river is the Mississippi River that runs from the Gulf of Mexico to the Lake Itasca in Minnesota. The Mississippi is a major transportation waterway through the middle of the US and has major tributaries, (Ohio and Missouri rivers) and is the third largest river system in the world. The total length of the Mississippi River is 3,763 km. The longest river in the US is the Missouri is 4,023 km.

A 4,827 km Intracoastal Waterway system provides protected waterway for commercial and recreational boats along the US Atlantic coast from Boston, Massachusetts to Key West, Florida (AIWW – Atlantic Intracoastal Waterway) and from Apalachee Bay, Florida to Brownsville, Texas (GIWW – Gulf Intracoastal Waterway) along the Gulf of Mexico coast. This free waterway was authorized by the US Congress in 1919 and is maintained by the US Army Corps of Engineers at a depth of 4 m. Commercial usage of the waterway includes barges hauling petroleum products, food, building materials, and manufactured goods. Annual dredging of the Intracoastal Waterway system is coordinated by the US Army Corps of Engineers.

6.8 US OPTIONS FOR PLACEMENT OF DREDGED MATERIAL

Guidelines for the disposal, or placement, of dredged material were developed through a joint effort of the Environmental Protection Agency and the US Army Corps of Engineers (3). These guidelines satisfy the National Environmental Policy Act, Clean Water

Act, and Marine Protection, Research and Sanctuaries Act and state water quality regulations. The guidelines allow for three possibilities for the disposal of dredged material. These possibilities are open water disposal, confined disposal, and beneficial use. All three possibilities must be addressed and a tiered approach is used to evaluate an acceptable disposal method. Testing procedures are defined in joint publications known as the Ocean Testing Manual (1) and the Inland and Near-coastal Testing Manual (2).

6.8.1 Open water placement

Placement of dredged material in rivers, lakes, estuaries and oceans is called open water disposal. Open water placement is accomplished using pipeline dredges, hopper dredges, barges/scows loaded by mechanical or hydraulic dredges, and mechanical dredges. The open water placement sites are described as nondispersive or dispersive sites. Nondispersive sites imply that the dredged material remains within the disposal site boundaries. Dispersive sites means the dredged material is expected to leave the disposal site due to environmental forces such as ocean currents or other meteorological events (e.g., storms, frontal passages, hurricanes, etc.). Cutter suction pipeline dredges discharge slurry into the water body and the solids eventually settle to the bottom forming a mound of dredged material. Some of the finer material may remain suspended and move out of the disposal area with the prevailing currents. Hopper dredges store the excavated dredged material in a large hopper within the vessel. Overflowing can increase the amount of sediment in the hopper. The overflow procedure allows more sand to settle in the hopper while the water and fine material overflows back into the water body. When the hopper is full, the hopper dredge sails to the open water placement site and opens the bottom doors, or opens its hull to allow the dredged material to fall through the water column and form a mound of material on the water body floor. Most of the dredged material comes to rest on the bottom. Mechanical dredges remove the dredged material at nearly in-situ density and place it in a barge or scow. Once the barge/scow is filled, a vessel tows the barge to the open water placement site where the dredged material exits the bottom of the barge/scow and falls to the bottom as shown in Figure 6.19. Open water dredged material placement sites are marked on navigational

Figure 6.19. Split hull hopper barge placing dredged material in open water (Courtesy of US Army Corps of Engineers).

charts and are typically found just outside ship channels where frequent dredging occurs. Rivers, bays, and estuaries also contain designated disposal sites.

6.8.2 Upland confined placement

Upland areas and near shore areas that are diked to contain dredged material and allow clear water to return to the water body are called confined disposal facilities (CDF) and are illustrated in Figure 6.20. Guidelines for the design, operation, and management of CDFs are contained in a USACE document (20). Cutter suction pipeline dredges are typically used to deliver the dredged material in the form of slurry through a long pipeline to the confined disposal site (Figure 6.21). The dredged material enters the CDF at an inlet structure and the dredged material flows through the site and ponds (Figure 6.21). An outlet structure (e.g., weir) holds back the slurry for sufficient time to allow the solids to settle. Once the water clarifies as a result of the sediment settling, then the weir boards are removed and the water overflows back to the water body (e.g., river, bay, etc.). Eventually, all the water leaves the site through drainage and evaporation and the dredged material dries and consolidates. Mechanical dredges sometimes load dredged material in barges/scows that are brought to the confined disposal site and unloaded onto trucks that are taken to the disposal site for

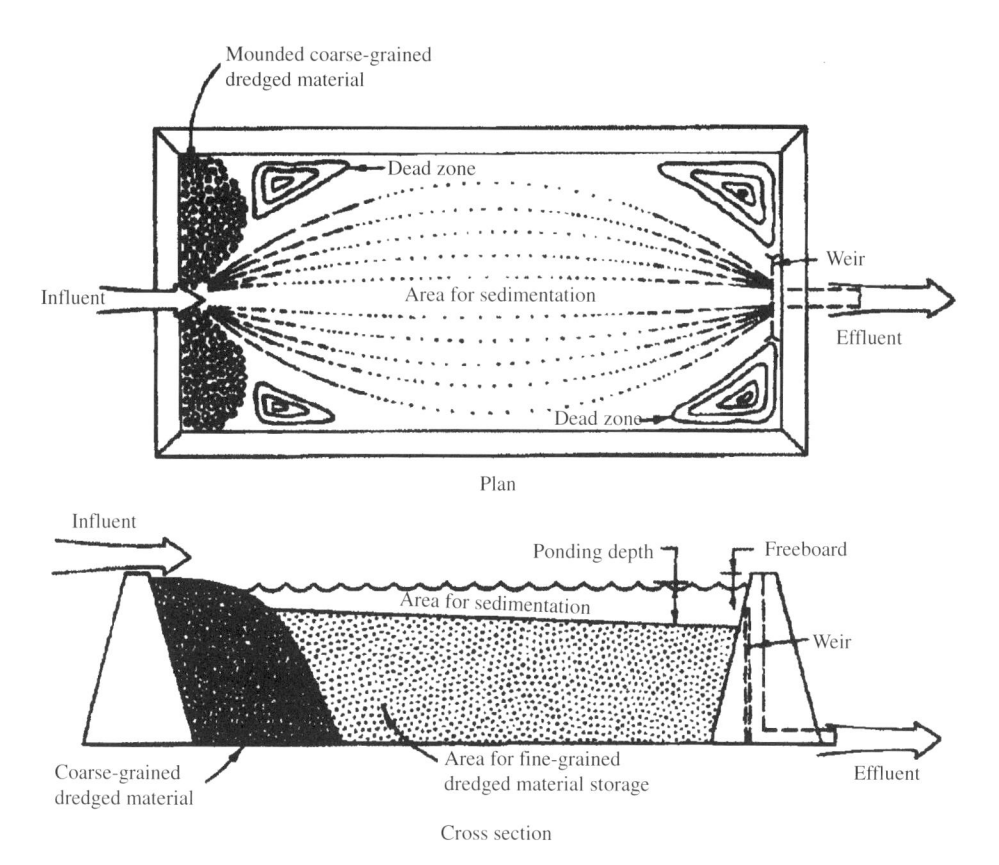

Figure 6.20. Schematic of a confined disposal facility (Courtesy of US Army Corps of Engineers).

Figure 6.21. Confined disposal facility after initial filling inlet (left) and after the filling (right) (Courtesy of Army Corps of Engineers.)

emptying or the dredged material is hydraulically slurried and pumped to the disposal site via a pipeline.

6.8.3 Beneficial use

Landin (21) describes in detail the many beneficial uses of dredged material. The US Army Corps of Engineers programs such as the Dredged Material Research Program (DMRP) from 1973–1978, the Dredging Operations Technical Support Program (DOTS) from 1978-present, the Environment Effects of Dredging Program (EEDP) fro 1982-present, the Dredging Research Program (DRP) from 1991–1996, the Dredging Operations and Environmental Research Program (DOER) from 1998-present, and the Wetlands Research Program (WRP) from 1990–1995 have contributed significant research results related to the beneficial use of dredged material. The Permanent International Association of Navigation Congresses (PIANC) also published a practical guide (22) for beneficially using dredged material. Conventional placement of dredged material at open water and confined disposal sites continues to be the most economical placement procedures. However, there is strong desire to find more funding and ways to economically use dredged material in a beneficial way.

Dredged material contains sands, silts and clays, and these sediments can be considered a resource that can be used to benefit society. The Corps of Engineers manual on beneficial uses of dredged material (23) describes ten broad categories of beneficial uses of dredged materials as habitat restoration, beach nourishment, aquaculture, recreation, agriculture, land reclamation and landfill cover, shoreline erosion control, industrial use, material transfer for dikes, levees, parking lots, highways, and multiple purposes. Using dredged material beneficially is very desirable, but the cost of bringing the dredged material to the location where it can be used beneficially is often a difficult and costly endeavor. Examples of beach nourishment, habitat creation, and agriculture reuse are illustrated in Figure 6.22.

Graalum, Randall, and Edge (24) suggest a potential beneficial use of dredged material is to use it to manufacture topsoil. Previous manufactured soil projects include Toledo Harbor, New York/New Jersey Harbor, Mobile Harbor in Alabama, and the Herbert Hoover Dike surrounding Lake Okeechobee in Florida. The manufactured soil helps reduce and recycle wastewater sludge and provides an additional alternative for the long-term management of dredged disposal sites by reducing the amount of land needed for the confined disposal facilities. Manufactured soil is created using dredged material and recyclable organic waste

After   Before
Beach nourishment   Habitat creation

Figure 6.22. Examples of beneficial uses of dredged material (Courtesy of US Army Corps of Engineers.)

materials. The organic waste materials may be bio-solids (e.g., sewage sludge, animal manure, yard waste), or biomass (e.g., cellulose or saw dust). The physical properties of the dredged material such as sediment grain size and composition are measured to determine the most appropriate use of the manufactured soil.

6.9   MODELING OF DREDGING AND DREDGED MATERIAL DISPOSAL

In the United States, an open water disposal site (ODMDS) is designated by the Environmental Protection Agency, and the site is managed by the US Army Corps of Engineers. Managing the site includes recording all disposal activities and monitoring the post disposal fate of the placed dredged material. The site manager must ensure that the dredged material does not accumulate in a manner that poses a navigational hazard, that unacceptable adverse impacts do not occur, and optimum volumetric utilization of the site is attained.

Numerical models have been developed to assist in the management of open water disposal sites. The first model is the short term fate of dredged material disposed in open water (STFATE) and it is described by Johnson (25 and 26). The STFATE model accounts for the physical processes that govern the immediate fate of dredged material disposed at open water sites typically over a time period ranging from minutes to hours. Among other features, it simulates the initial bathymetric deposition pattern and associated thickness of the dredged material from as single disposal event. The resultant sediment distribution on the receiving bathymetry is represented according to a local coordinate grid format. Separate algorithms account for the method (hopper or barge) by which the dredged material is released into the open water. STFATE is not capable of adding the results of multiple disposal cycles that are typical of large disposal operations.

Once dredged material has been placed at the site, it forms a mound-like feature for each individual placement. Multiple placements form a composite mound and these newly created mound structures are exposed to the physical processes governing the erosion and consolidation of the dredged material within the ocean environment. After the composite mound is in place, then the long term fate model (LTFATE) can be applied. This model simulates the morphology of a dredged material mound through coupled hydrodynamic and sediment transport equations and includes the effects of subaqueous consolidation over time periods from weeks to years (27). MDFATE incorporates modified versions of STFATE and LTFATE to predict the bathymetry resulting from a series of disposal cycles and the period following the disposal project activity when environmental forces act on the open water dredged material disposal site (ODMDS) (28 and 29).

For upland disposal in diked confined disposal facilities (CDF), models developed through the Corps of Engineers include SETTLE and PSDDF. SETTLE uses laboratory settling test data to determine the required size of a CDF for solids retention and initial storage of the dredged material and it evaluates the weir length required for prevention of excessive discharge of suspended solids. PSDDF is a consolidation program used in the long term management of confined disposal sites and has been used to determine the level of sediment after consolidation for wetland creation projects. These software programs are available from the Corps of Engineers (*www.wes.army.mil/el*).

6.10 DREDGING INFORMATION, RESEARCH AND EDUCATION

Dredging research in the United States is conducted by US Army Corps of Engineers (USACE) (*www.usace.army.mil*) and several academic institutions. The major dredging research program being conducted by the Corps of Engineers is Dredging Operations and Environment Research Program (DOER) (*www.wes.army.mil/el/dots/doer*). This large multiyear research program is managed by the US Army's Engineering Research and Development Center (ERDC) (*www.erdc.usace.army.mil*) in Vicksburg, Mississippi (formerly known as the Waterways Experiment Station). The Center consists of several laboratories and the two most involved with dredging research are the Coastal and Hydraulics Laboratory (*http://chl.wes.army.mil*) and the Environmental Labora-tory (*www.wes.army.mil/el*). Dredging research reports can be accessed through DOER web site.

The Western Dredging Association (WEDA) (*www.westerndredging.org*) is a professional society that serves the dredging community in the western hemisphere by disseminating dredging research through annual technical conferences and published proceedings. WEDA holds an annual Technical Conference and it is one of three organizations that make up the World Dredging Organization (WODA). A World Dredging Congress is conducted every three years and is hosted by one of the three WODA members, WEDA, CEDA (Central Dredging Association), or EADA (Eastern Dredging Association). The American Society of Civil Engineers (ASCE) (*www.asce.org*) sponsors dredging conferences occasionally and the last conference was Dredging '02. There is also the US Section of the Permanent International Association of Navigation Congresses (PIANC) (*www.iwr.usace.army.mil/PIANC/*) that sponsors conferences related to navigation dredging.

The Center for Dredging Studies (*http://edge.tamu.edu*) in the Ocean Engineering Program at Texas A&M University in College Station, Texas conducts an annual Texas A&M Dredging Seminar. The one day Texas A&M Dredging Seminar is currently held in conjunction with the annual WEDA Technical Conference. Dredging short courses are conducted by the Center every year in January on the campus of Texas A&M University. Additionally, a graduate level course on Marine Dredging is offered every two years.

Test facilities for test dredge pumps and modeling dredging operations and dredged material placement are found at the GIW Industries Laboratories (*www.giwindustries.com*)

in Groveton, Georgia, the Coastal and Hydraulics Laboratory in Vicksburg, Mississippi, and the Coastal Engineering Laboratory (*http://edge.tamu.edu*) at Texas A&M University.

There are two dredging magazines published in the United States. These near monthly magazines are the World Dredging, Mining, and Construction magazine and International Dredging Review magazine. The Western Dredging Association publishes the Journal of Dredging Engineering quarterly.

ACKNOWLEDGEMENTS

The author acknowledges the support of Ocean Engineering Program in the Civil Engineering Department at Texas A&M University during the writing of this chapter. The assistance of graduate student Robert Adair and undergraduate student Samantha Saunders in generating some of the figures is greatly appreciated.

REFERENCES

1. USEPA/USACE. "Evaluation of Dredged Material Proposed for Ocean Disposal (Testing Manual)." EPA-503/8-91/001, Office of Water, US Environmental Protection Agency, Washington, 1991.
2. USEPA/USACE. "Evaluation of Dredged Material Proposed for Discharge in Inland and Near-Coastal Waters – Testing Manual." In preparation, Office of Water, US Environmental Protection Agency, Washington, 1996.
3. USEPA/USACE. "Evaluating Environmental effects of Dredged Material Management Alternatives – A Technical Framework." EPA 842-B-92-008, Washington: US Government Printing Office, November 1992.
4. Herbich, J. B. *Handbook of Dredging Engineering*. Second Edition, New York: McGraw-Hill, 2000.
5. Reine, K. J., Dickerson, D. D., and Clarke, D. G., "Environmental Windows Associated with Dredging Operations," DOER Technical Notes Collection (TNDOER-E2), U.S. Army Engineer Research and Development Center, Vicksburg, MS, 1998.
6. Marine Board, "A Process for Setting, Managing and Monitoring Environmental Windows for Dredging Projects," Transportation Research Board and Ocean Studies Board, National Research Council, March 19–20, 2001.
7. Anonymous, "Cabotage and the U.S. Economy: Employment, Transportation and the National Interest". 5-30-03. *http://www.mctf.com/Fact-iii.htm*.
8. Anonymous, "Lake Carriers' Association – Truth versus Fiction about the Jones Act". 5-30-03. *http://www.lcaships.com/jones.html*.
9. Anonymous, "The Jones Act: Fact and Fiction". 5-30-2003. *http://www.mctf.com/jonesact.htm*.
10. Anonymous, "Vital Role of Cabotage in Maintaining Competition in Domestic Shipping: Debunking the Deregulation Myth". 5-30-03. *http://www.mctf.com/Fact-iv.htm*.
11. Anonymous, *The Jones Act – Does It Still Make Sense?* The National Waterways Conference, Inc. Seminar Proceedings. Arlington, Virginia. November 1989.
12. Sickles, Mark. "Letter to the Editor – Jones Act – a Reprise". *WORLD DREDGING Mining & Construction*. p.19, 21. January 2003.
13. *World Dredging, Mining and Construction (WDMC)*, Dredging Equipment Directory, March 2003.
14. *International Dredging Review (IDR)*, Directory of Dredge Equipment, April 2003.
15. Marine Board, "Contaminated Sediments in Ports and Waterways," National Research Council, National Academy Press, Washington, DC, 1997.

16. Palermo, M. R., Clausner, J. E., Rollings, M. P., Williams, G. L., Myers, T. E., Fredette, T. J., and Randall, R. E. "Guidance for Subaqueous Dredged Material Capping", Technical Report DOER-1, US Army Corps of Engineers, Washington, DC, June 1998.
17. Palermo, M. R., Maynord, S., Miller, J., and Reible, D. D. "Guidance for In-situ Subaqueous Capping of Contaminated Sediments," Environmental Protection Agency, EPA 905-B96-004, Great Lakes National Program Office, Chicago, IL, 1996.
18. Science Applications International Corporation (SAIC), "Sediment Capping of Subaqueous Dredged Material Disposal Mounds: An Overview of the New England Experience, 1979–1993," DAMOS Contribution #95, SAIC Report No. SAIC-90/7573&c84 prepared for the U.S. Army Engineer Division, New England, Waltham, MA, 1995.
19. Standing Committee on Water Transportation, "A Report on Water Transportation Dredging and Dredged Material Management: An Intermodal Issue," American Association of State Highway and Transportation Officials: Washington, DC, 1998.
20. USACE, "Confined Disposal of Dredged Material", Engineer Manual, US Army Corps of Engineers, EM-1110-2-5027, September 1987a.
21. Landin, M. L. "Beneficial Uses of Dredged Material" Chapter 16, In *Handbook of Dredging Engineering*, Second Edition, Herbich, J. B. Editor, New York: McGraw-Hill, 2000.
22. PIANC, Beneficial Uses of Dredged Material: A Practical Guide, Report of PIANC Working Group 19, Permanent International Association of Navigation Congresses (PIANC), Brussels, Belgium, 1992.
23. USACE, "Beneficial Uses of Dredged Material", Engineer Manual, US Army Corps of Engineers, EM-1110-2-5026, June 1987b.
24. Graalum, S. J., Randall, R. E., and Edge, B. L. "Methodology for Manufacturing Topsoil Using Sediment Dredged from the Texas Gulf Intracoastal Waterway," *Journal of Marine Environmental Engineering*, Vol. 5, pp. 121–158, 1999.
25. Johnson, B. H. "User's Guide for Models of Dredged Material Disposal in Open Water", Environmental Laboratory, Technical Report TR D-90-5, US Army Engineer Waterways Experiment Station, Vicksburg, MS, 1990.
26. Johnson, B. H. "Development and Verification of Numerical Models for Predicting the Initial Fate of Dredged Material Disposal in Open Water, Report 2: Theoretical Developments and Verification", Environmental Laboratory, Technical Report DRP-93-1, US Army Engineer Waterways Experiment Station, Vicksburg, MS, 1995.
27. Scheffner, N. W., Thevenot, M. M., Tallent, J. R., and Mason, J. M. "LFTATE: A Model to Investigate Fate and Stability of Dredged Material Disposal Sites", Technical Report DRP 94-6, Coastal Engineering Research center, U.S. Army Waterways Experiment Station, Vicksburg, MS, 1995.
28. Moritz, H. R. "A Coupled Strategy for Numerical Modeling the Fate of Dredged Material Placed in Open Water", Dredging Engineering Short Course Notes, Center for Dredging Studies, Texas A&M University, College Station, Texas, January 2000.
29. Moritz, H. R., and Randall, R. E. "Simulating Dredged Material Placement at Open Water Disposal Sites", *Journal of Waterway, Port, Coastal, and Ocean Engineering*, Vol. 121, No. 1, 1995.

Chapter 7

Dredging in Hong Kong

Peter G.D.Whiteside
Independent Consultant, formerly Chief Geotechnical Engineer, Geotechnical Engineering Office, Government of the Hong Kong Special Administrative Region, People's Republic of China

7.1 INTRODUCTION

More than 400 million cubic metres of material dredged in four years and more than 100,000 cubic metres of trailer dredger capacity operating in the middle of one of the world's busiest ports – these impressive headline statistics from the 1990s marked the key role played by modern dredging in the most intensive period of development Hong Kong has ever seen.

Hong Kong lies on the southern coast of China at the eastern edge of the estuary of the Pearl River, which is China's third largest river (Figure 7.1). Despite being described

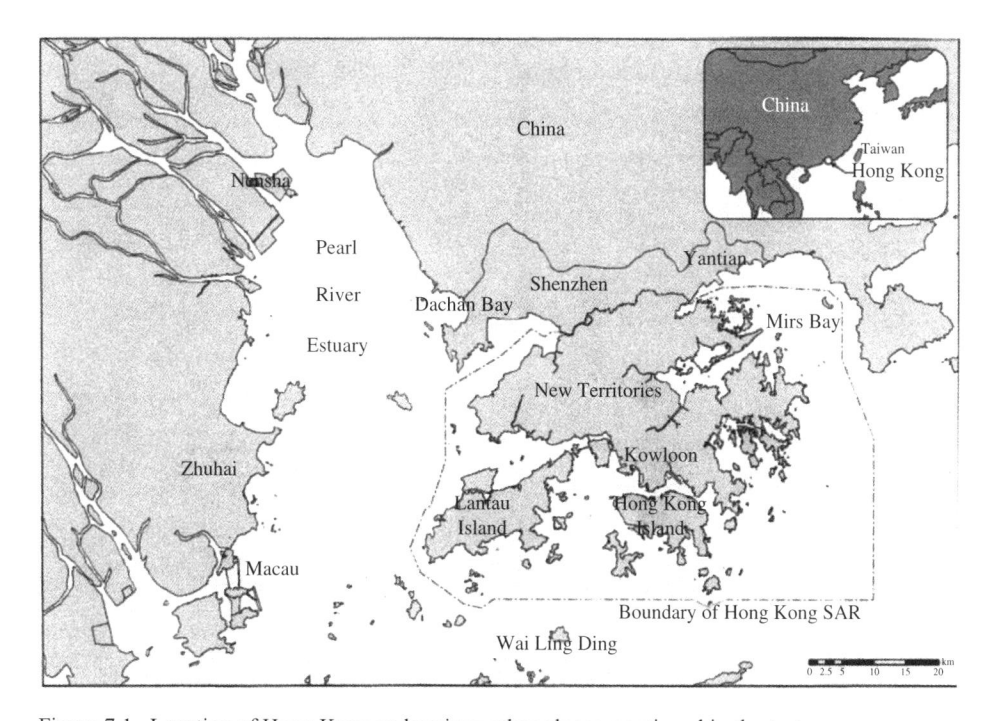

Figure 7.1. Location of Hong Kong and various other places mentioned in the text.

a century and half ago as a "barren rock", Hong Kong has become world-renowned as a financial, business and above all, trading centre. The British administration of Hong Kong ceased on July 1st, 1997 when Hong Kong became a Special Administrative Region (SAR) of the People's Republic of China, but Hong Kong's key position as an entrepot to southern China remains unchanged and is still centred around its fine deep-water port.

Although Hong Kong comprises some 1,000 square kilometres of land, two thirds of this is either hilly terrain which is too steep to develop, or small isolated islands. The population, which at the end of 2002 was over 6.8 million, is therefore concentrated in highly developed urban areas with an average population density of more than 6,000 people per square kilometre. Although new towns have been established in many parts of Hong Kong during the last fifty years, the historical concentration of population around the harbour area is still evident today.

Hong Kong is the world's 10th largest trading economy and has one of the world's largest banking centres, stock markets, foreign exchange and gold markets, and is a major exporter of trade-related and financial services. However, the port continues to be the key factor in the prosperity and economic growth of Hong Kong. Annually it handles about 200 million tonnes of cargo or 80 per cent of Hong Kong's total cargo throughput, with an average of over 200,000 ocean-going and coastal vessel movements. For most of the last decade, the container port has been the world's largest and today it has an annual throughput of about 20 million TEUs (Twenty foot Equivalent Units). In 2004, six new container berths come on stream adding an annual capacity for a further 2.6 million TEUs.

Hong Kong is also a major international and regional aviation centre and the recently completed new airport when fully developed will have an annual capacity for nearly 90 million passengers and 9 million tonnes of cargo.

The subject of dredging in Hong Kong is covered below under the following broad headings:

- Hong Kong's natural marine environment
- Reasons for dredging in Hong Kong
- Environmental regulatory framework
- Sand dredging
- Mud dredging and disposal
- Future dredging in Hong Kong

Although the environmental regulatory framework is separately described, specific environmental issues relating to dredging works are discussed within the sections on sand dredging, and mud dredging and disposal.

7.2 HONG KONG'S NATURAL MARINE ENVIRONMENT

Hong Kong's mountainous landscape is mostly composed of granite and volcanic bedrock. Deep weathering and erosion have produced material which has been transported and then deposited in the valleys as brownish alluvial clays, silts and sand. Global sea-level changes associated with ice ages have repeatedly drowned this rugged landscape – the most recent and highest relative sea-level having been reached around 6,000 years ago during the Holocene marine transgression. Onshore, the valleys remain but offshore they have been covered by the sea leaving the steep sided hills as islands (Styles & Hansen, 1989; Fyfe, et al., 2000). The rising Holocene sea caused extensive erosion of the steep hillslopes and the fine component of the material so liberated was deposited as a swathe of mud covering most of the new marine area. The present-day seabed can be divided into two types of zone: along the coast, a sandy and rocky sloping seabed which is now colonised by an abundant and diverse fauna including molluscs, crustaceans and corals, and over most of the remaining sea area, a flat muddy seabed dominated by marine worms with lesser numbers of molluscs and crustaceans. The activity of natural anaerobic organisms within this muddy sediment has given it a dark grey colour.

The present-day bathymetry of Hong Kong waters increases from northwest to southeast. It is generally less than ten metres in the northwest, where Hong Kong borders the Pearl River estuary, becoming ten to twenty metres in the central harbour area, and then reaching about thirty metres in southeastern waters. Tidal flows are the dominant influence in Hong Kong's inshore area with current speeds in excess of 2 m/s in constricted channels but 0.5 m/s or less in more sheltered waters. The pattern of tidal currents is very complex but the strongest currents and residual currents are in a generally southeast-northwest direction and this is reflected in the bathymetry which features two main tidal channel networks which are fifteen to twenty metres deeper than the surrounding seabed and which snake their way through constricted island gaps. One of these networks of tidal channels passes through the central waters of Hong Kong and provides the main deep shipping access to Hong Kong's container port.

Two seasonal currents affect the offshore waters of Hong Kong. The summer months, or wet season, are dominated by a current flowing in a northeasterly direction with associated warm monsoon winds from the southwest, while in the winter the dominant current flows in a southwesterly direction with associated cold monsoon winds from the northeast. The transition between the two is characterized by periods of unstable and unpredictable current directions. The tidal currents in the inshore waters are affected by these seasonal offshore currents, but in the wet season, the effect of the Pearl River discharge is also very pronounced. In these months, the river's discharge is greatest and penetrates into western and central Hong Kong waters as a low salinity surface current that can also carry a high level of suspended sediment. Eastern Hong Kong waters are far less influenced by the Pearl River. They tend to be very clear, and for this reason most of the popular beaches and diving sites are to be found along the eastern coastline.

During the summer months, cold oceanic water with a lower oxygen content wells up from the continental shelf in the south and penetrates the lower layer of inshore waters. The extent of the resulting stratification varies from year to year but can be enhanced by the warmer surface water flow from the Pearl River estuary.

The normal wave climate in Hong Kong waters is characterized by calm conditions (<0.3 m wave height) for about 90% of the time and a gentle swell (<0.6 m) coming in from the southeast for the remainder of the time (Ip, 1996). There are, however, two exceptions to these conditions. During the months June to October, tropical cyclones are regular features in the South China Sea and they often impact Hong Kong. Onshore, the intense rainfall and high winds pose serious problems of landslips and flooding while offshore, apart from the disruption of normal maritime traffic, the principal effects are the greater wave height and the associated disturbance to the seabed sediment. The other occasion when higher waves can be expected is during the prolonged periods of strong winter monsoon when a heavy swell from the northeast can affect smaller vessels and can also cause significant remobilization of seabed sediment. These two exceptions are critically important from an environmental point of view because during these periods of poor weather, water quality sampling boats are not normally deployed and so routine datasets of water quality tend to miss the natural events of high suspended sediment and therefore give an incomplete picture of the natural conditions under which the marine ecosystem exists. This is discussed later in more detail.

7.3 ENVIRONMENTAL REGULATORY FRAMEWORK

7.3.1 Classification and disposal of dredged sediment

Pollution control policy in Hong Kong is developed and implemented by the Environmental Protection Department (EPD) and for many years Hong Kong has been implementing controlling legislation to reduce water pollution at source. Nevertheless, the marine sediments close to developed areas are still contaminated by past pollution. Over many decades, both domestic and industrially polluted wastes have found their way via foulwater sewers, stormwater drains and watercourses, into Hong Kong's marine environment. Extensive field sampling and laboratory testing have revealed large amounts of metallic and organic pollutants. Not surprisingly, concentrations have tended to be highest around submarine outfall pipes. Once on the seabed, pollutants have then been mixed deeper into the sediment by natural processes but also by ships' anchors. As a result, contamination is commonly present in the seabed sediment to depths of up to 3 m and occasionally more. However, much of this contaminated seabed sediment has now been removed and disposed of as part of the major dredging works of the 1990s.

Hong Kong's "Dumping at Sea Ordinance", originally stemming from UK legislation, follows the London Convention (LC) in implementing measures for the avoidance of marine pollution, in particular, the licensing and monitoring of disposal operations. Because China is also a signatory to the LC, this situation did not change after the 1997 reunification of Hong Kong with China. Unlike many other jurisdictions, however, Hong Kong's "baseline" for the purposes of the LC-compliant local regulations is the highwater mark, therefore all dredging activities, even small backhoe operations within tidally affected freshwater drainage channels, are covered by the same legislation.

In 1989, EPD adopted interim guidelines for categorisation of severely contaminated dredged material in the northwest part of Hong Kong. In 1992, EPD replaced these guidelines with a scheme for classification of dredged material that was based on the levels of contamination by seven heavy metals (Cd, Cr, Cu, Hg, Ni, Pb & Zn). This scheme was used to discriminate between material which would be suitable for disposal at one of the open sea disposal sites and material which would require isolation from the environment.

In the years following 1992, chemical testing of sediment to be dredged expanded beyond the seven heavy metal contaminants, and biological effects testing was examined with a view to developing a more comprehensive regulatory approach. About this time, the LC adopted a new framework for assessing dredged material (IMO, 1995) and this approach was adopted in Hong Kong. In 1996, an independent strategic review of the sediment testing,

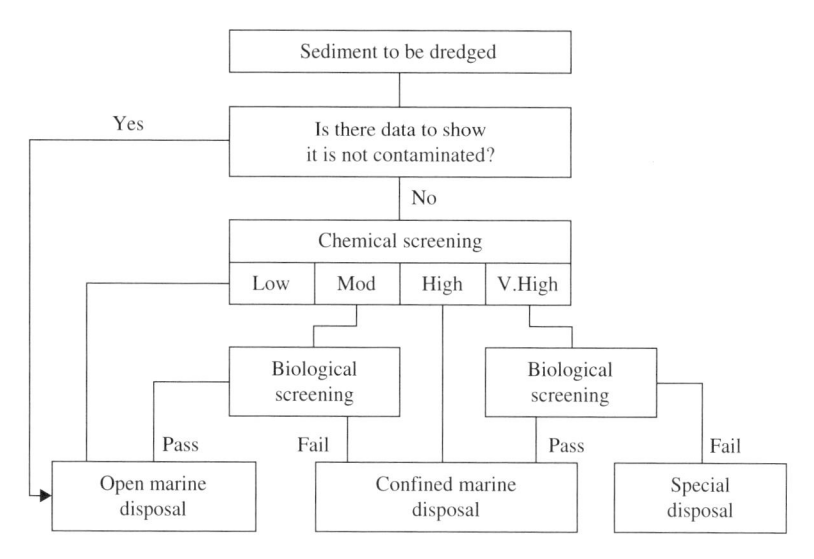

Figure 7.2. Scheme for testing and classification of dredged sediment for disposal.

classification and disposal arrangements was undertaken (EVS, 1996a & b). Amongst other things, this review identified the possibilities for incorporating biological effects testing into routine testing. By 1998, EPD had identified the outline for a new scheme of classification (Lei et al., 1998), and the new regulations were promulgated in April 2000 and became mandatory for all dredging works which commenced from the start of 2002 onwards. Full details of these regulations are given in the Hong Kong government's Environment, Transport and Works Bureau Technical Circular No. 34/2002 "Management of Dredged/Excavated Sediment", but the essential stages in testing and classification of dredged sediment for disposal are illustrated in Figure 7.2 adapted from Whiteside et al. (2001).

7.3.2 Environmental Impact Assessment

Projects which have a potential to adversely affect the environment are required to undergo the Environmental Impact Assessment (EIA) process. Prior to 1998, the regulations made provision for an initial Environmental Review to determine the likely extent of the impacts and to determine whether or not a full EIA would be required. In 1998, the "Environmental Impact Assessment Ordinance" came into force to avoid, minimise and control environmental impacts through the EIA process and through the Environmental Permit. Reference should be made to the EIA Ordinance for full details, but in essence all projects designated under a Schedule to the Ordinance are required to undertake an EIA and then, after approval of the EIA by the Director of Environmental Protection who takes advice from amongst others the non-governmental Advisory Council on the Environment, to apply for and then adhere to the conditions of an Environmental Permit. EIAs are publicly displayed and arrangements are made for the views of all interested parties to be considered. EIAs will normally incorporate a programme of Environmental Monitoring and Audit (EM&A),

the results of which are critically important in demonstrating compliance with the legally binding Environmental Permits.

As will be described later, many of the sand dredging operations were covered by the pre-1998 regulations but nevertheless were still governed by stringent EM&A programmes. A key aspect of EIAs for dredging works has been undertaking computerised hydraulic modelling to predict the extent of suspended sediment, nutrient and contaminant dispersal. In the early 1980s the government established a mainframe mounted, 2-Dimensional, depth-averaged model for studying the hydraulic and pollution dispersion impacts of proposed marine works. Later, as computing power increased and as dredging impacts, *per se*, attracted more consideration, the government established a 2-Dimensional, two layer wet and dry season model supported by an extensive programme of field calibration (Rodger, 1996). The capability of the model to study suspended sediment dispersion using random walk plume modelling or similar was an important component of both the government modelling programmes and those run by independent consultants. Progress in modelling was rapid as small desktop computers became the norm and as hydraulic models became multilayered and more sophisticated in general.

Separately from the EIA Ordinance, the "Water Pollution Control Ordinance" established Water Control Zones for Hong Kong's marine waters. There are currently ten such zones and for each zone, specific Water Quality Objectives (WQO) are established. With respect to suspended sediments, the WQO for all zones is that waste discharges shall neither cause the ambient level to be raised by 30% nor give rise to accumulations of suspended solids which may adversely affect aquatic communities. The ambient level is not, however, defined in the Ordinance and EM&A programmes normally attempt to provide sufficient information to enable a judgement to be made.

Another important law in Hong Kong relating, *inter alia*, to proposals for dredging works is the "Foreshore & Seabed (Reclamations) Ordinance" which requires publication of any proposals for major dredging works so that any interested parties can lodge an objection. All the areas for sand dredging in Hong Kong were subject to the provision of this law.

7.4 REASONS FOR DREDGING IN HONG KONG

7.4.1 Sand to form new land as reclamations

From the very earliest years, Hong Kong has had a need for additional land for its expanding trading activities and population. The wharves and warehouses of the early traders quickly occupied the thin strip of flattish coastal land on Hong Kong Island and the steepness of the natural terrain left two options for further development on the Island: either build uphill on the steeper slopes or construct reclamations by infilling marine areas along the coastal strip (Tregear & Berry, 1959). In the event, both options were followed but it was the formation of reclamations which offered the greater potential for providing new land for development both on Hong Kong Island and across the harbour in Kowloon. Over the last century and a half more than 60 square kilometres of new land has been added to Hong Kong's total area in this way.

Hong Kong's shoreline has therefore moved progressively seaward by filling. Over the hundred years from 1860 to 1960, filling took place at a rate that averaged about 1 million

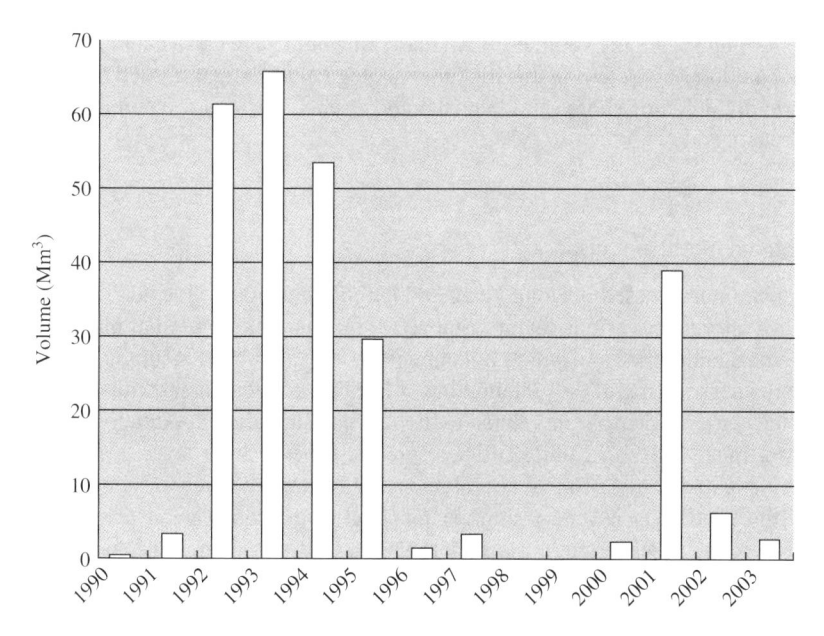

Figure 7.3. Annual requirements for dredged sand in Hong Kong 1990–2003.

cubic metres per year. Thereafter, the rate of land development increased significantly to satisfy the demands for space for housing, industry and infrastructure. Not only were there more reclamations formed in the harbour area but new towns were established in the flatter-lying areas beyond the central harbour. During the twenty years up to the mid-1980s the rate of reclamation filling increased to an average of some 15 million cubic metres per year (Brand & Whiteside, 1990). Thereafter, even greater volumes of fill material were required for the massive reclamation programme of the first half of the1990s when an average of nearly 70 million cubic metres of general and rock fill were placed per year. Figure 7.3 shows the annual volumes of dredged sand fill used since 1990. The period between 2000 and 2003 saw another peak in reclamation filling but the scale of reclamation currently planned beyond that is much less.

In the early days, almost all of the reclamation fill material came from excavation into hillslopes and the complete removal of small hills – a process that worked well because the excavated land also provided sites for development. However, the topography and geology of Hong Kong limited the scope for this process of sourcing material. The bedrock in Hong Kong's main developed area is dominated by deeply weathered granite which in the early days provided a mix of completely decomposed material which was used as general fill material and good rock which could be used for breakwaters and seawalls. By the 1980s, however, there were few land areas left which could be developed into such convenient borrow areas (an exception being the new airport site described later). But perhaps most importantly, environmental concerns in the 1980s and 1990s argued strongly against the development of massive land-based borrow areas. Quite apart from the destruction of the natural hillsides and their habitats, and the creation of permanent scars on Hong Kong's landscape, the short-term aspects of dust and traffic impacts from excavating, processing and transporting such huge volumes of fill material would have had a major adverse impact on the population. It was during this time that geological investigations led to the discovery of hitherto unknown offshore sand deposits and opened the way to sourcing reclamation fill by dredging.

7.4.2 Mud removal and disposal

As reclamation progressed in Hong Kong and harbour development extended further from the original shoreline, various factors changed – factors that are predominantly related to the geological conditions and which fundamentally affect both the sourcing of fill material and the civil engineering of the reclamations. Although greater water depths are part of this picture, the main differences are related to the seabed foundation conditions. As previously mentioned, most of Hong Kong's offshore area is covered by a layer of soft, dark grey marine mud. Close to the original shoreline and in constricted areas where tidal flows are stronger this mud layer is thin or non-existent, but away from the shore the thickness of the mud layer increases to ten or more metres. This soft marine mud has been extensively dredged for four main reasons.

- **Reclamation foundations:** Although some Hong Kong reclamations have been constructed on top of seabed mud deposits, it has been more common to remove all the soft mud deposits not just for foundations of seawalls and breakwaters but also from entire reclamations so as to reduce both the degree and the duration of consolidation settlement.
- **Navigation purposes:** Considerable quantities of mud have been dredged to deepen existing navigation channels and anchorage areas and to provide new ones.
- **Removal of mud overburden before sand extraction:** Because almost all of the offshore sand resources lie beneath the layer of soft mud, this mud overburden has to be removed prior to sand extraction.
- **Maintenance dredging:** Hong Kong was first established as a trading port because the waters were deep and sheltered. Most of the Pearl River's massive sediment load is actually carried down the western side of the estuary past Macau and there is no noticeable natural sedimentation in Hong Kong (Chalmers, 1983). Artificially deepened areas, however, and areas near watercourse outfalls do accumulate material that requires removal. Historically, the annual maintenance dredging requirement has been about 0.5 million cubic metres which is relatively small for a port the size of Hong Kong. This requirement gradually increased during the 1990s and was likely to average about 1.5 million cubic metres over the four years from 2001 to 2004. Most of the maintenance dredging is government funded but that for the container port and approach channels to the power stations is privately organised.

Figure 7.4 shows the annual volumes of mud dredged since 1990. As can be seen from the figure, about 10% of the dredged mud was categorised as contaminated and requiring special disposal. As will be described later, the special facilities for disposal of contaminated mud involved isolating the material in seabed pits – a process which itself resulted in a further category of mud dredging, namely that to form and cap the special disposal pits.

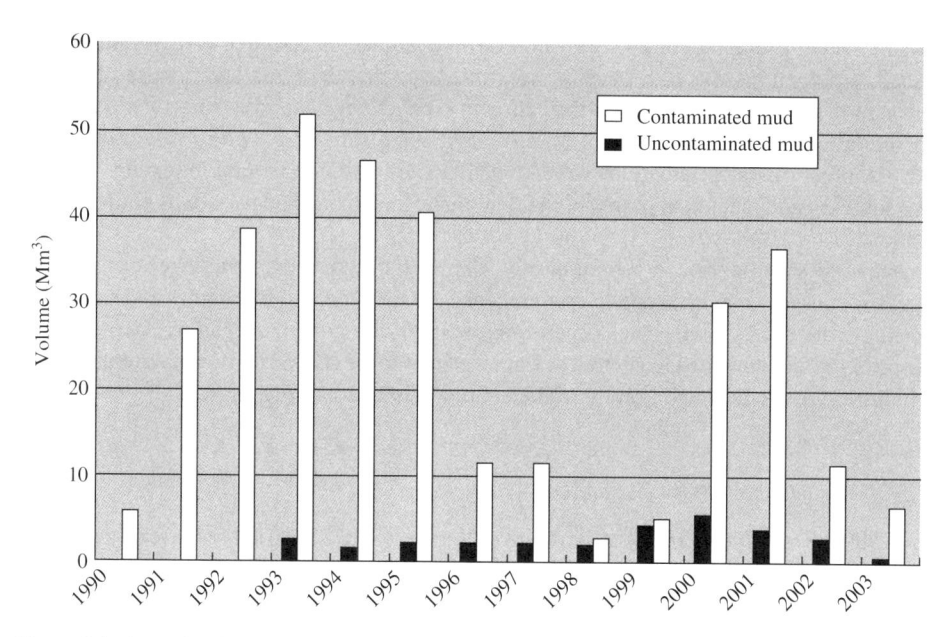

Figure 7.4. Annual volumes of mud requiring dredging in Hong Kong 1990–2003.

7.5 SAND DREDGING

7.5.1 Historical perspective

Marine granular deposits have been worked for many years in Hong Kong but it was only after 1950 that the volumes became very significant. Sources were generally near-shore areas where hydrodynamic conditions left coarse material exposed on the seabed. Sand was won by grab dredgers, bucket dredgers and small cutter suction dredgers. Although in some cases small amounts were won using local knowledge and little in the way of specialised field investigations, larger projects were preceded by comprehensive marine investigations using combinations of simple probing and marine boreholes. In the 1950s, 7.5 million cubic metres of sand was dredged for the reclamation to extend the runway of the now disused Kai Tak Airport. Although the grab-dredged material had more than 25% fines, the subsequent re-handling by cutter suction dredgers which pumped the material into settling areas in the reclamation left sand with less than 5% fines after simple overflowing drainage (Guilford, 1988). Several million more cubic metres of sand were dredged for other projects in the following years and by similar types of dredging plant. Environmental factors were of some concern in the 1950s but unlike the present-day, it was the noise from large bucket and grab dredgers in the central harbour area that was perhaps of greatest concern.

Despite these previous uses of marine fill in Hong Kong, the general lack of sand on the surface of Hong Kong's seabed led to a perception in the early 1980s that the marine sand option had perhaps been exhausted. This picture changed dramatically towards the end of the 1980s when the pattern of the offshore geology began to be understood and when Hong Kong began to better appreciate the immense possibilities offered by trailer suction hopper dredgers and other modern dredging plant. The experience brought to Hong Kong in 1986 by the Dutch dredging contractors who built the Container Terminal 6 reclamation with locally-dredged marine sand gave the clear message that dredging and the use of marine fill was the way forward for land formation in Hong Kong. This was very timely because in the late 1980s planning was at an advanced stage for Hong Kong's new airport and several other major reclamations which together were going to require in just one decade, the sourcing and placing of more fill material than Hong Kong had used in its entire history (Brand et al., 1994).

Although some marine investigations for fill material were being undertaken on a project-specific basis, it was clear that a comprehensive understanding of Hong Kong's offshore geology was of key importance. Therefore, in 1987, the government's then Geotechnical Control Office, now the Geotechnical Engineering Office (GEO) started a systematic programme of geological exploration of the entire offshore area of Hong Kong.

7.5.2 Management of sand resources

By 1989, although the GEO's exploration programme had identified significant resources of sand it was clear that without efficient and coordinated management of these resources there would not be sufficient material for all reclamation contracts at the times it was required. The government decided that a centralised system would be required to manage the sand resources and it set up the multidisciplinary, interdepartmental Fill Management Committee (FMC), now termed the Marine Fill Committee. The responsibilities of the then FMC were soon expanded to include policy responsibility for management of mud disposal. Technical and administrative support for the Committee is provided by the GEO's Fill Management division which prepares policy proposals; implements policy decisions; maintains a database of projects' dredging and other requirements; maintains a geographic information system of all dredging and disposal areas and related data; and which also undertakes the technical studies, including EIAs, related to sand dredging and mud disposal. Almost all information is in the public domain, much of it on-line at http://www.cedd.gov.hk/eng/services/fillmanagement/index.htm where project data, statistics, maps, lists of reports, etc. are available. The establishment of the FMC was an important and necessary departure from previous government administrative arrangements, and the adoption of this holistic approach, whereby the planners and engineers for all reclamations were brought into one policy forum together with environmental, maritime and other statutory regulators, proved to be an efficient way of accommodating all interests and enabling timely decisions to be made.

7.5.3 Exploration for sand and examples of some major dredging projects

An impetus to the early stages of the GEO's sand exploration programme was the independent identification and later the extraction of large quantities of marine sand for two big land formation projects. Before continuing the discussion of the GEO exploration it is worth digressing to give a brief account of these two projects because both projects highlighted important aspects of sand dredging which had to be taken into account in the GEO exploration programme to ensure that economic, dredgable fill resources were identified and not just geological deposits.

Container Terminal 6 dredging project: The first project, already mentioned above, was the reclamation for Container Terminal 6 in the central harbour area. A detailed account of this project is given by Wragge–Morley (1988). Work commenced with removal of the soft seabed mud in the reclamation area. Despite seabed debris and an uneven surface at the base of the soft mud deposits, some 7 million cubic metres of seabed mud were removed from the reclamation site in just five months using two large trailer suction hopper dredgers. A third trailer dredger was then mobilised to assist with the filling operation using a deposit of shelly seabed sand that had previously been identified only five kilometres from the reclamation site. After just four months, 8 million cubic metres of marine sand (>90% of the total required) had been dredged and placed – 5 million by bottom discharge and 3 million by pumping either through a bow nozzle or via a floating pipeline. A medium-sized cutter suction dredger was then mobilised to assist in pumping fill material to parts of the reclamation not accessible to the trailers and to form surcharge embankments. A bucket dredger and some barge-mounted grabs were also used for slope trimming work. Quite apart from the unique ability of the trailer dredgers to complete such an enormous volume of work in such a short time – sand filling rates peaked at more than one million cubic metres per week – they demonstrated the vital importance of having dredging plant that is mobile. To have attempted such dredging works with static plant would have imposed severe restrictions and risks on the movement of shipping in this very busy area of Hong Kong's harbour. However, the use of trailer dredgers with state-of-the-art navigation and positioning systems, permitted safe working 24 hours a day, seven days a week. Indeed, in many later dredging contracts in Hong Kong, the use of such mobile trailer dredgers became a specified requirement. Although environmental concerns received rather less publicity in the late 1980s than they were to receive in Hong Kong during the 1990s, environmental monitoring was undertaken during the sand dredging. The overflow process, so essential to the economic filling and operation of trailer hopper dredgers, also resulted in a dramatic reduction of fines content of the sand delivered to site compared with the fines content of the sand in-situ. This had two obvious effects: first, overflow plumes were more intense and second, more seabed material could meet the specification for filling material than had been estimated from the original particle size analysis of the seabed deposits – in fact, nearly twice the volume of useable fill material anticipated at the project planning stage was recovered. While the additional volume of usable material was of great importance it was perhaps not unexpected to experienced dredging contractors. The results of the environmental monitoring of suspended sediment released from overflowing provided another positive aspect, namely that despite the high visibility and intensity of the overflow plumes, the suspended sediment was found to have settled rapidly and the associated impacts both to the water column and to the surrounding seabed were considered to have been localised and short-term (Holmes, 1988). Furthermore, the environmental study of the works suggested that the recolonisation of affected areas of the seabed, although not monitored, would probably occur shortly after cessation of dredging. From an environmental perspective, the lack of significant noise pollution associated with larger trailer dredger operations was also an important factor.

Tin Shui Wai New Town dredging project: Immediately after formation of the reclamation for Container Terminal 6, land formation works started for Tin Shui Wai New Town, planning of which included the need for 24 million cubic metres of sand to infill a low-lying coastal area in the northwest of Hong Kong. The ground investigation for the project had identified a number of sand deposits all of which were covered by a thickness of marine mud. One large sand deposit, located some way further down the coast from the reclamation site, was found to have a relatively thin layer of mud overburden. The geology and the delineation of all these sand deposits was somewhat more complicated than in the Container Terminal 6 case and the detailed nature of the site investigation and geological interpretation pointed the way for future investigations (Dutton, 1987). The contractor for the works used a trailer dredger for removal of mud overburden and then the sand was extracted in a series of pits using a dust-pan dredger loading barges, the pits later being available for disposal of mud. The barges transported the sand about 15 kilometres to where a very large cutter suction dredger, assisted by a booster pumping station, pumped the sand several kilometres to the land formation site. It is interesting to note that in a nearby marine area, the contractor also experimented with a production-sized airlift system pumping compressed air down the annulus of a double pipe so as to extract sand from beneath thicker overburden without having to remove the overburden. The system worked but it is understood that the sand's coarseness and angularity restricted its ability to flow towards the riser pipe and so it was abandoned. In the Tin Shui Wai project, as with the Container Terminal 6 project, the responsibility for identifying and managing the use of the sand resource lay with those controlling the project. In the event, this proved satisfactory in the sense of overall management of Hong Kong marine fill resources because in both these cases effectively the entire sand source was used for each project and the question of how and when to access remaining sand for other projects did not arise. With the larger and more extensive resources discovered and later utilised during the dredging of the 1990s this was not to be the case and the FMC's central management of resources became very important.

Knowledge from these two projects was invaluable to the GEO's sand exploration programme. The initial stage of the GEO programme was development of a geological model of the present-day offshore area. As discussed earlier, the present-day offshore area of Hong Kong was land during the sea level lows associated with the ice ages and during this time sand was deposited along the valley river courses while the floodplains were sites of silt and clay deposition. The rising sea then inundated the valleys and deposited a layer of soft mud. The initial geological model was therefore of ancient river channel sands buried beneath a swath of soft mud.

The site investigations were carried out in two distinct phases (Whiteside & Massey, 1992) and one of the critically important factors was extensive local experience in undertaking high resolution marine geophysical surveys (Ridley Thomas et al., 1988). The first stage of fieldwork consisted of an approximately 3 km grid of seismic survey using a 1 kHz model frequency boomer with swell filter. The seismic survey was complemented by a series of vibrocores and continuously sampled boreholes. This first stage investigation established that the geological model was over-simplified. Two aspects were particularly relevant to the sand search. Firstly, was the fact that during the early stages of the marine transgression when shallow tidal seas swept over the pre-Holocene land surface, winnowing of the earlier sediments had left thin sheets of sand (Shaw, 1988; Whiteside, 1991; Martin et al., 1997) which in certain places had accumulated into low profile, migrating sand banks (Evans, 1988). Comminuted shells are a common component in these sands and in some locations constitute a high proportion of the material. Secondly, the layer of Holocene mud that buried the sand and other deposits was shown by the seismic surveys to be thin or even absent in constricted areas between islands where tidal flows were greater (Arthurton, 1987). The geological model was therefore revised and selected sand bodies targeted for more detailed

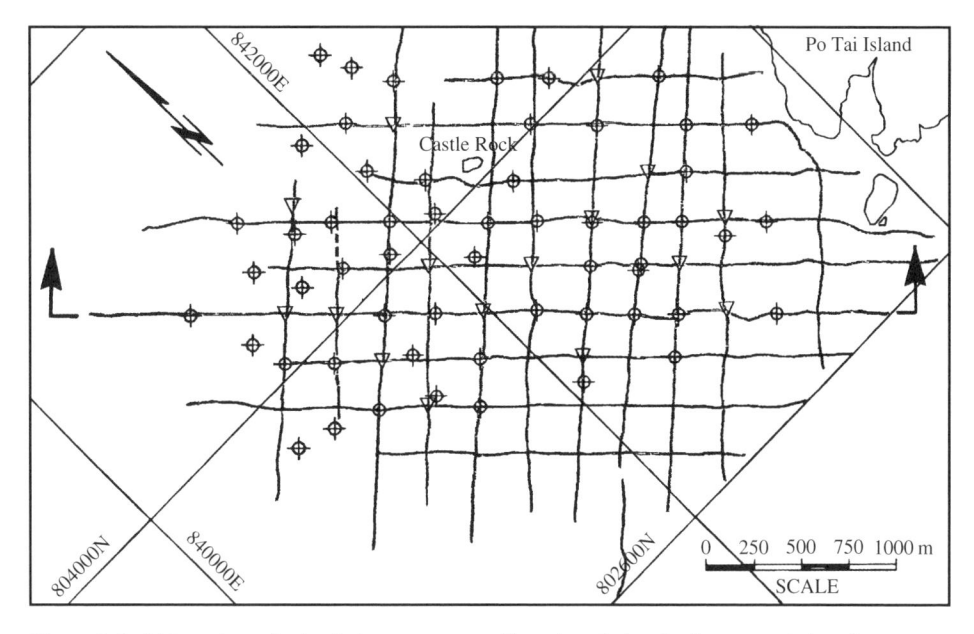

Figure 7.5. (a)Location of seismic boomer survey lines, boreholes & vibrocores (⊕) and cone penetration tests (∇) at West Po Toi (after DEMAS, 1996) Line of Figure 7.5(b) cross-section also shown.

investigation. It is worth noting that the sand sources dredged for Container Terminal 6 and for Tin Shui Wai New Town were both located in present-day tidal channels. The Container Terminal 6 sand source was an early Holocene sandbank with a very high shell content and almost no mud cover while the Tin Shui Wai sand source was mostly pre-Holocene river sand with a relatively thin cover of mud.

The second stage of the GEO investigation comprised further boomer surveys, but this time on an approximately 350 m grid accompanied by piezo-cone penetration tests (CPTs) and cable tool boreholes – the latter permitted continuous sampling of all layers and enabled Standard Penetration Tests (SPTs) to be carried out (Binnie Consultants Ltd, 1991–93). The use of quick and inexpensive CPTs (about 30% of the cost of a borehole) proved very successful. On average about ten borehole/CPT stations were used per square kilometre. Synthesis of the seismic, borehole and CPT data was critical since the digitised interpreted seismic sections were then used to construct a gridded computer model of the area for reserve calculation and borrow area development design and planning. Borehole and other investigation data such as SPT values and particle size determinations, much of it supplied on diskettes in a pre-specified format by the site investigation contractors, were stored on a Geographic Information System (GIS) using ARC/INFO software (Selwood & Whiteside, 1992) and then exported to a personal computer. Sediment layer levels and dredging parameters were then calculated at the nodes of the chosen grid so that volumes, stripping ratios, etc. could be calculated for each grid square (e.g., DEMAS, 1996). Figures 7.5(a) and 7.5(b) show respectively, the plan layout of site investigation at West Po Toi and a cross-section through the borrow area. Defined economic reserves were then used by the FMC to allocate dredging areas to individual projects. An essentially similar, though much more

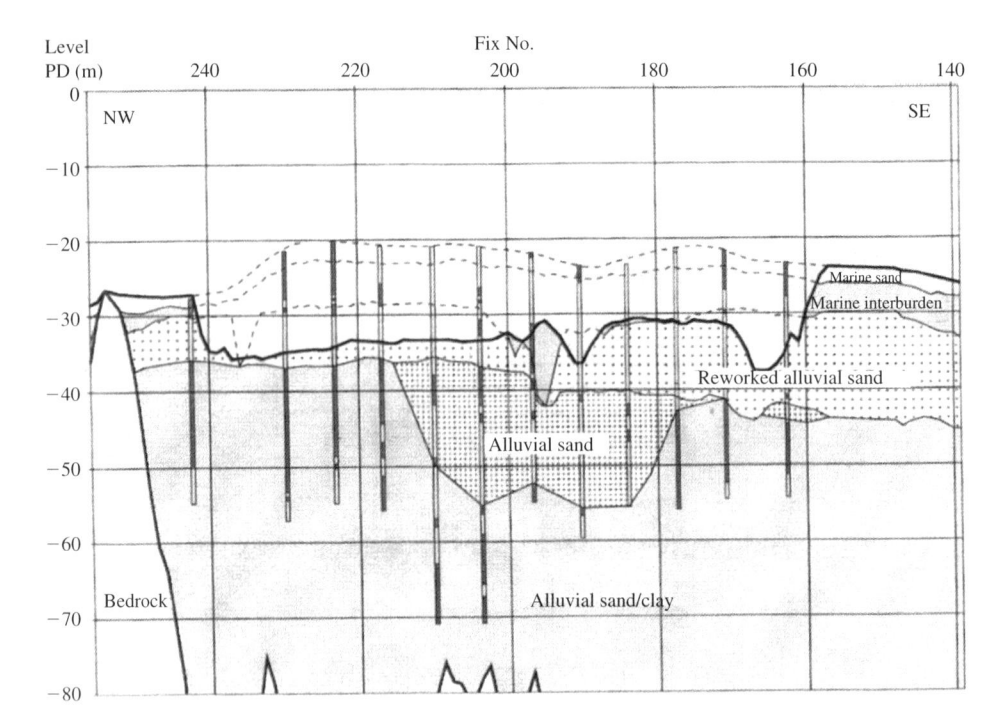

Figure 7.5(b). Cross-section through West Po Toi marine borrow area showing borehole/vibrocore/CPT locations and interpreted seabed sediment layers (DEMAS, 1996).

detailed process of reserve calculation was used by the dredging contractors in planning and controlling of the actual extraction. An important Government decision at the time was to make both factual and interpretative borrow area reports available to tenderers for dredging contracts so as to assist them in making comprehensive assessments of borrow areas. This move was generally welcomed by dredging contractors and project engineers.

Most of the offshore sands dredged for use as fill material are alluvial quartz sands derived predominantly from nearby weathered granitic bedrock. The upper layers of many deposits contain a small proportion of comminuted shell fragments, and lithic fragments are common in deposits derived from terrain which includes volcanic and other rocks. The sands range from fine to coarse but with little gravel fraction – the size of the quartz grains being dependent on the coarseness of the parent bedrock and, as with the lithic grains, on the distance from the material source. Distance of sediment transport is also the main control on grain angularity and sphericity, although some of the upper layers of the alluvial sands which were tidally reworked by the early Holocene sea can be well rounded. However, many of the alluvial sand deposits have not been transported great distances and the sub-angular quartz grains can be very abrasive to dredging plant. This is particularly important when sand is pumped significant distances such as in the Tin Shui Wai project where sand was pumped several kilometres.

The various types of dredging plant and methods employed for sand winning and placing in the Container Terminal 6 project have already been described and a similar pattern was to be maintained throughout the intensive dredging activities of the 1990s. However, because the sand deposits in borrow areas mostly extended to depths not normally attainable by trailer dredgers, different techniques were used to extend the reach of many of the dredgers so as to gain as much sand as possible from a borrow area. As a result, most borrow areas could be dredged to depths of between 35 and 40 m, and in one case, dredging for Container Terminal 8, the use of a submersible pump part-way down the suction pipe enabled sand to be dredged from a depth of 54 m (de Kok, 1994). In the marine borrow areas, sand winning, preceded by removal of mud overburden, was almost entirely undertaken by trailer dredgers while cutter suction dredgers with pipeline networks assisted in placing the sand at the reclamation sites. Although cutter suction dredgers were also used in some cases for mud removal and sand winning – a small cutter was used in the Soko Islands borrow area south of Lantau Island, for example – it was the medium to large size trailer suction hopper dredger that was the key workhorse of all the large projects. Indeed, in the early 1990s at the peak of dredging activity, a sizeable proportion of the world's medium to large trailer dredgers with a combined hopper capacity in excess of 100,000 cubic metres were working in Hong Kong on sand and mud dredging (Figure 7.6). For these few years Hong Kong became a world centre of dredging work (Ooms et al., 1994) and a description of the largest project, the new airport, is central to any account of dredging in Hong Kong.

New airport and related dredging projects: The new airport was constructed on a reclamation adjacent to Lantau Island and involved the levelling and extension of a small island called Chek Lap Kok. The levelling produced over 100 million cubic metres of mixed soil and rock fill, with the balance of the fill material being dredged sand (Malone & Oakervee, 1993; Plant et al., 1998). Because hitherto there had been no fixed transportation route between the new airport site and the mainland of Hong Kong, bridges and land formation were needed to convey the new road and rail links. Major new reclamations using dredged sand were required along the coasts of North Lantau and Kowloon. And in the central harbour area between Kowloon and Hong Kong Island, large volumes of mud had to be dredged for the construction of new submersed tube road and rail links. These links extended from Central Reclamation via the West Kowloon Reclamation and the North Lantau Expressway to the new airport as shown on Figure 7.6. Figure 7.6 also illustrates the complex arrangement of different marine borrow areas supplying sand to these projects and to Container Terminal 8 which was under construction at the same time. Many dredgers alternated between sand and mud dredging and considerable use was made of the opportunities for combined cycling whereby the four activities comprising mud removal at the reclamation site, mud disposal at the designated marine area, sand winning at the marine borrow area, and sand placement at the reclamation site were combined into one dredging cycle (Vlasblom, 1994). With most of the major dredging contractors represented in all the main projects and contracts, it was also not surprising that individual trailer dredgers also alternated between different projects when programming permitted. Side contracts for beach replenishment were also undertaken. The largest single reclamation contract, however, was the airport's Site Preparation Contract (SPC). The SPC involved some 76 million cubic metres of dredged sand and 106 million cubic metres of dredged mud (Plant et al., 1998). The SPC's joint venture contractor and its sub-contractors commenced mud dredging in early 1993, initially to remove soft deposits at the reclamation site and later to remove overburden at the marine borrow areas. By about mid-1993 the mud dredging rate had peaked and sand winning rates were increasing. Between, mid-1993 and mid-1994, an average of about 12 million cubic metres of material was being dredged each month.

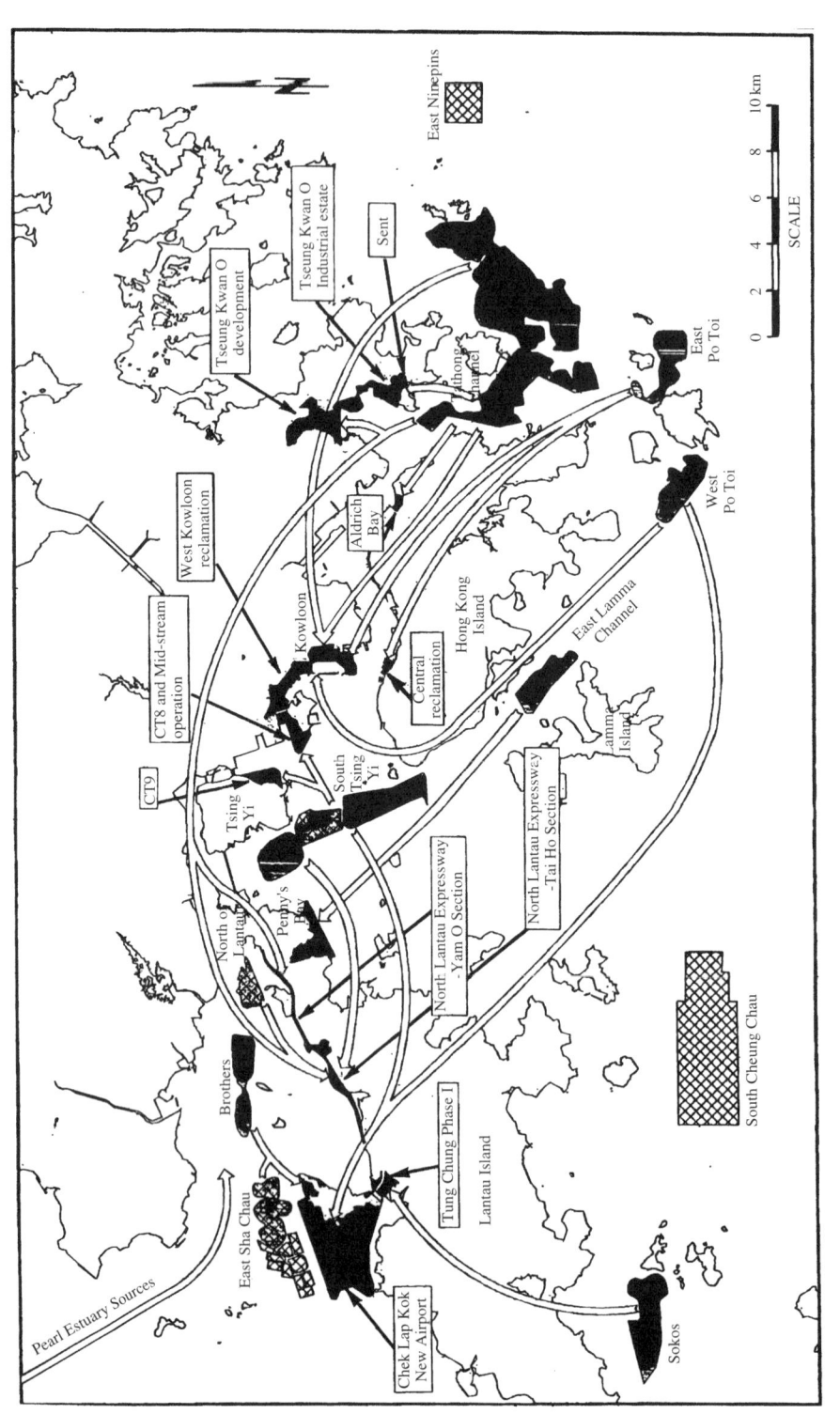

Figure 7.6. Location of major reclamations (black), marine borrow areas (shaded) and mud disposal sites (hatched). The arrowed lines show which sand source was used for which reclamation and all were active during the early 1990s except Container Terminal 9 (CT9) and Penny's Bay for which the sand was dredged between 2000 and 2002.

Dredging was essentially complete by mid-1995 and Plant et al. (1998) reported that at the peak of activity, some fifteen trailer dredgers, five cutter suction dredgers, seven grab and bucket dredgers as well as numerous barges and tugs were working on the contract. Capacity of grab dredgers ranged from 6 to 13 cubic metres while that of their companion barges varied from 700 to 1,000 cubic metres.

7.5.4 Fines content of sand: implications for fill specification and dredging

Kwan (1993) reviewed the specifications used in Hong Kong for dredged marine fill and a number of inconsistencies and problems were identified. Specifications for sand fill designed to be used in similar engineering situations ranged from <10% fines content to <35% fines with a tendency for faster tracked more critical contracts to be in the range <10% fines to <20%. However, almost all the sand was trailer-dredged, so achieving the specification posed no problems because the overflow during hopper loading reduced the fines content significantly. This meant that although the *in situ* fines content of the various sand deposits varied from <35% to near zero, any marine borrow area would be suitable for almost any contract no matter what fill specification was used. An exception, however, arose with some very small contracts that used grab-dredging to extract the sand because this method resulted in very little fines loss. Despite the fines loss in larger contracts using trailer dredgers, dredging contractors still had to carefully control fill rehandling and placement by pipeline so as to avoid problems with segregation of the remaining fines.

Most of the sand dredged has been alluvial sand originally laid down along the courses of rivers – a process which means deposits are finer grained the further they are from the original material source. There is therefore a tendency for the sand in the more distant borrow areas to be finer and to have a higher fines content. This factor makes it difficult to optimise the loading of the hopper so as to achieve the maximum rate of sand delivery at the reclamation site. This is because the greater the sailing times between borrow area and reclamation site the more important it becomes to load the hopper as fully as possible, however, the finer the material, the more difficult and time consuming it is to fill the last portion of the hopper (Whiteside et al., 1998). In at least one of the more distant borrow areas the long overflow times resulted in siltation of the sand dredging site which had to be cleared and the silt taken for disposal.

7.5.5 Environmental aspects of sand dredging

In the very early stages of the GEO's exploration for dredgable sand, some deposits in southeastern waters were rejected for further study because they were located just offshore from sandy beaches and their removal would almost certainly have resulted in sand loss from the beaches during storms. In another area, Deep Bay in northwestern waters, a sand deposit was also rejected for further study because it was known that the overburden mud was contaminated. Other locations which were identified as potential marine borrow areas underwent the non-statutory environmental assessment process (i.e., this pre-dated the 1998 EIA Ordinance). As explained earlier, the process required an initial desktop Environmental Review study of each area to determine the likely extent of impacts. In two cases, EPD decided that further, detailed EIA studies were required. One of the EIAs undertaken covered the large sand deposits in Mirs Bay in eastern waters. The EIA concluded that

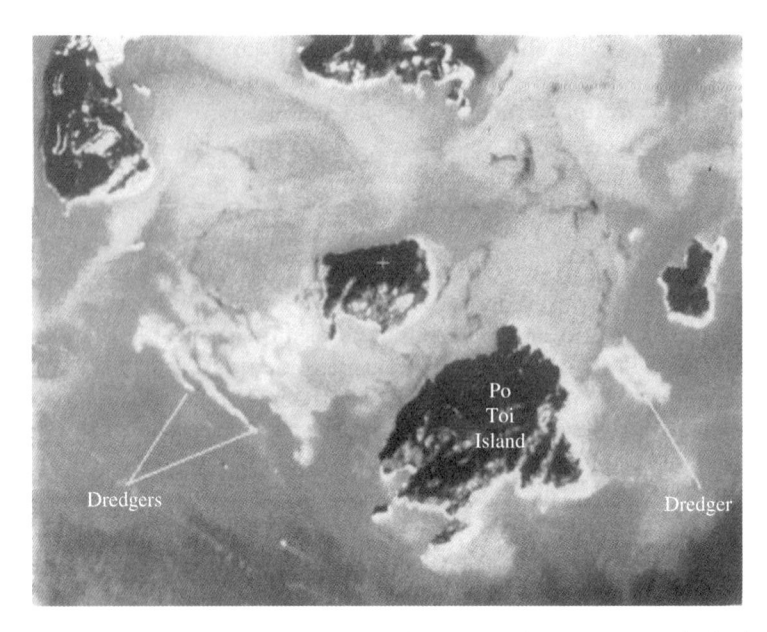

Figure 7.7. Satellite photo of suspended sediment plumes from dredging at Po Toi Island.

despite the possibility of incorporating mitigatory measures in any dredging operation, it was likely that there would be residual environmental impacts on corals and other marine life (Binnie Consultants Ltd, 1992), and on this basis the FMC decided not to proceed and the proposal was cancelled. The second EIA related to a sand deposit in the East Lamma Channel, immediately west of Hong Kong Island (Binnie Consultants Ltd, 1993). In this case the EIA indicated that mitigatory measures could reduce impacts to acceptable levels and the FMC decided to proceed with the proposal, although it was to be many years before the area was used. Eventually the sand was dredged for the reclamation in Penny's Bay to provide land for the Hong Kong Disneyland Resort (Penny's Bay reclamation in Figure 7.6).

Although the early dredging works attracted some public attention, concerns increased as more sand dredging work got underway. The concerns comprised not only the general public's intuitive reactions to plumes of turbid water – they look dirty and hence are polluting – but also the more considered worries of informed green groups and some academics that widespread dispersion and subsequent settling of suspended sediment could irrevocably damage areas of Hong Kong's marine ecosystem. Despite the body of knowledge on this subject from elsewhere in the world, these concerns had to be addressed in detail in the Hong Kong context and the series of surveys and studies which were undertaken by the GEO and its consultants concentrated firstly on providing scientific data on the physical and ecological nature of the local marine environment, and secondly on the measurement and analysis of the generation, decay and fate of actual sediment plumes. Figure 7.7 is a satellite photo of dredging plumes around the Po Toi Islands in southeastern waters where many detailed studies were undertaken. The area is also shown in Figure 7.6. The investigative work to examine the public's concerns took several years to demonstrate that these impacts are indeed limited to the immediate area of dredging, and that the biological activity on the seabed restores the disturbed ecosystem, in most cases in a relatively short time. The

GEO's environmental monitoring has been widely reported as has the overall conclusion that dredging impacts are both transient and localised (Evans, 1994; Ng et al., 1998; Ng & Chan, 2004). The programme of monitoring work was diverse in the techniques employed and physically extensive but only a very brief summary is given below. In essence the programme examined physical and ecological effects and concentrated on three things : the muddy flattish seabed; the rocky coastal seabed; and individual dredging plumes.

On the muddy seabed, grab sampling, faunal analysis, and innovative seabed camera surveys demonstrated that the seabed conditions are naturally very dynamic and that the ecosystem coped with all but the most intense sedimentation that occurred immediately adjacent to borrow areas – and even in these areas, colonisation by burrowing deposit feeders started taking place immediately after cessation of dredging (Germano et al., 2002). By the end of the 1990s, independent studies by prominent academics were also beginning to demonstrate that dredging impacts to the marine ecosystem were short-lived and localised and that intensive trawler fishing was actually a more damaging activity (Leung & Morton, 2000).

On the rocky coastal seabed, extensive dive surveys (Binnie Consultants Ltd, 1995b) identified some areas immediately adjacent to sand dredging where intensive sedimentation had smothered some soft corals and some hard table corals. Dive surveys of other areas where dredging-related elevations of suspended sediment levels had been recorded indicated, however, that high levels of suspended sediment did not *per se* result in coral mortality. In 1994, during the period of the dive surveys, the normal summer stratification of eastern waters became very intense. This was probably related to a 1 in 100 year rainfall event in southern China, with a consequently high freshwater discharge from the Pearl River, coupled with the lack of major storms that would have mixed the different layers. The colder, oxygen-poor water layer, rose to within two metres of the surface and resulted in mortality of corals and other marine fauna over an area of some 200 square kilometres (Binnie Consultants Ltd, 1995a; Evans, 1994). While this event did not mitigate the localised dredging-related damage, it did help put coral mortalities into the context of natural impacts. In some areas close to corals, mitigation measures were taken involving various restrictions on the sand dredging (and mud backfilling of the borrow pit when empty – see later in this section). Some restrictions were tide-related and essentially involved only operating parts of the area close to corals when the tidal currents would carry suspended sediment away from the coral. In some locations this was not straightforward because tidal current directions were influenced by the strength of the monsoon winds. Elsewhere, mitigation to avoid damage to coral included restricting the number of dredgers and the weekly sand production until the EM&A had demonstrated that rates could be increased.

Individual dredging plumes were studied intensively using satellite and aerial photography and multi-boat field monitoring with water samplers, siltmeters and acoustic doppler current profilers – the latter had been developed in backscatter mode as suspended sediment meters (Land et al., 1994). Field results were also compared to computerised hydrodynamic modelling predictions which were made using the actual dredging and overflow parameters. Results of plume monitoring at West Po Toi marine borrow area have been reported elsewhere (Whiteside et al., 1996) but in essence they showed that in the first five to ten minutes after overflow, the behaviour of the sediment-water mixture is dynamic and the bulk of the material moves to the seabed as a density current. Part of this density current is, however, entrained in the water column and forms a plume of suspended sediment that gradually decays as individual particles settle under gravity. Depending on the characteristics of the sand being dredged and the method of dredging, the plume almost completely decays to background level after about two hours in a similar recent study of sand dredging plumes in Hong Kong, Cheung & Ho (2004) reported decay to background levels in one and a quarter hours in the East Lamma Channel marine borrow area. Studies in Hong Kong have also showed that computerised modelling of plumes can give reasonably reliable predictions of plume movement and decay, provided an appropriate particle settling velocity is used because the life of the plume, and hence the distance it can travel, is almost *pro-rata* the settling velocity. The studies in Hong Kong indicated that model predictions of sand dredging plumes can give reasonably reliable predictions with a settling velocity of 1 mm/sec (Whiteside & Rodger, 1996).

From the onset of field studies of the marine ecology and sand dredging plumes, knowledge of the natural background level of suspended sediment and its variability was vitally important because as previously mentioned, Hong Kong's Water Control Ordinance regulations target 30% over ambient as the normally acceptable limit. These regulations formed the basis of permissible levels included in EM&A programmes for dredging works but implementing such regulations can be very complex not least because there are many ways of comparing time or otherwise averaged ambient levels with the short-lived elevations in mixing zones around dredging works. Another difficulty related to determining acceptable levels of suspended sediment impact for marine fauna, especially corals and other fixed organisms. Prior to this time the main data on suspended sediment levels was from routine water quality monitoring stations located all around Hong Kong waters. However, as mentioned at the very beginning of this chapter, there are frequent though rather short duration natural events normally missed in routine monitoring. These events – pulses of very high sediment load from the Pearl River during the rainy season, suspensions of sediment by strong monsoon winds from the northeast, and tropical cyclones or typhoons – result in very significant elevations in suspended sediment levels, often considerably higher than those from dredging operations. By lucky chance, a typhoon passed through Hong Kong in 1995 at a time when continuously recording siltmeters had been installed on the seabed in the vicinity of the dredging operations shown in the satellite photo in Figure 7.7. Parry (2001) has reported on this in detail but in essence this data showed how insignificant dredging impacts are in comparison with natural variations in suspended sediment levels, and it also demonstrated, by inference, that short-term high levels of suspended sediment did not have long-term adverse effects on either Hong Kong's water quality or its marine fauna. Figure 7.8 shows suspended sediment levels near the dredging during the passing of the typhoon. Large amounts of other suspended sediment data were collected during the busy dredging years – perhaps the most notable being the deployment of acoustic doppler current profilers on survey boats which traversed all Hong Kong waters in a major field exercise (Dredging Research Ltd, 1995). By the year 2000, an enormous body of data had been accumulated on Hong Kong's natural patterns of suspended sediment and this data now helps provide greater confidence in assessing the short-term, localised impacts from many types of dredging activity.

The GEO's efforts concentrated on concerns at the sand dredging sites and mud disposal areas. At the reclamation sites there were major environmental concerns related to dispersion of fines from sand pumped onto reclamations and from mud dredging. In these cases the project's own team used their own EM&A programme and sometimes additional monitoring to help address the issues and to identify whether working methods needed to be

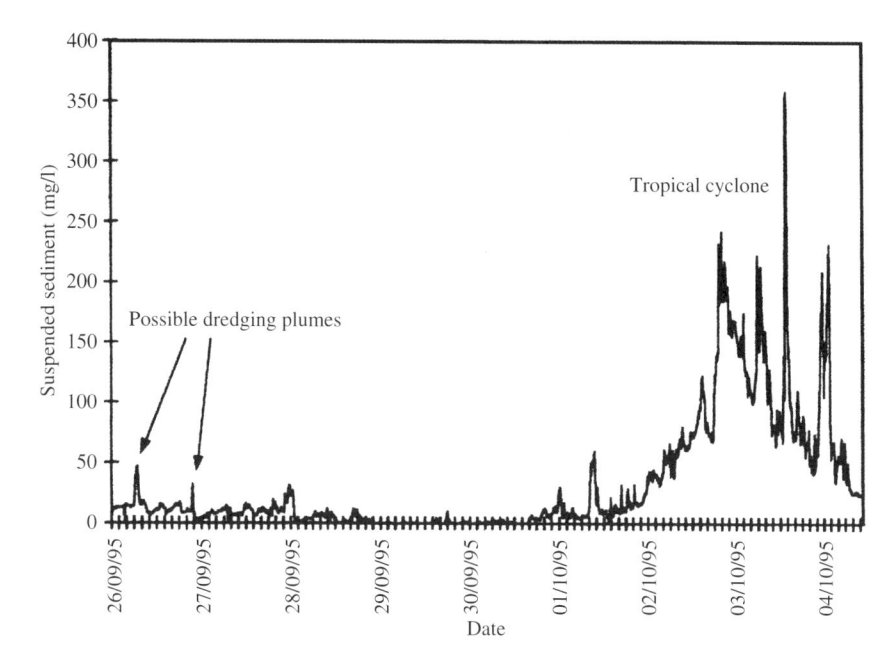

Figure 7.8. Turbidity levels near Po Toi dredging during the passage of a tropical cyclone.

adapted. Diverse issues such as potential ecological damage and commercial consequences of fish deaths in mariculture sites had to be addressed simultaneously – with the added complication that this was all occurring in a relatively small area where many other independent marine activities took place. Resolution of the issues, though in part aided by revision of dredging practices, placement of silt curtains, and so on, was often protracted and rarely straightforward.

Finally, under the heading of environmental aspects of sand dredging it is worth mentioning the FMC's programme to use empty sand borrow pits for disposal of mud. This was a long-term programme adopted not only because disposal capacity at open sea disposal mounds was limited but also because it provided an environmentally sound arrangement by which the dredged part of the seabed could eventually be restored to its original, natural state. The presence of artificially deep areas of seabed also affects the water flow characteristics and re-instatement of the original bathymetry serves to re-establish the local wave climate and general hydrodynamics. Furthermore, because almost all borrow areas were originally covered by a layer of seabed mud, backfilling an empty sand pit with mud reinstates a muddy seabed. From a marine environmental perspective, the sustainability of this process is evident. In contrast, however, extraction of the limited sand reserves is clearly not sustainable and so for some years now Hong Kong has been importing sand from the huge reserves in neighbouring parts of mainland Chinese waters.

7.5.6 Supply of dredged sand from outside Hong Kong

As Hong Kong's programme of reclamations moved from planning to reality in the 1990s, sand suppliers across the border in mainland China geared up to provide fill material, and despite the greater travel distance, the practice of sourcing sand from outside Hong Kong has been growing steadily ever since. Apart from the recent construction of Container Terminal 9 and the Hong Kong Disneyland Resort, which together used much of Hong Kong's last easily accessible sand reserves, the demand for sand fill in Hong Kong has been met by imported sand since late 1996. Sand extraction in mainland Chinese waters is subject to various regulations and procedures which involve official bodies with both regulatory and resource management roles. While contractors working on reclamation projects in Hong Kong do not have to pay any fee for use of Hong Kong marine borrow areas, they do have to pay a royalty for sand removed from mainland waters.

Even before the reunification of Hong Kong with China, cross-border liaison had resulted in Hong Kong's GEO undertaking marine sand exploration within neighbouring Chinese waters. Notable among the areas investigated was one immediately south of Hong Kong near Wai Ling Ding island (see Figure 7.1) where considerable volumes of sand were discovered. The sand in this location is finer but the thickness of mud overburden is less than commonly found in some of the near-shore borrow areas within Hong Kong. Although not as attractive a resource as the coarser deposits closer to Hong Kong project sites, this deposit has proved an effective borrow area and some 40 million cubic metres of sand has been extracted for use in construction of the Hong Kong Disneyland Resort at Penny's Bay on Lantau Island (see Figure 7.6). In this case it was the project contractor in Hong Kong whose trailer dredgers extracted and transported the sand to site.

Other sand sources in several of the municipalities in the Pearl River region have also been used to supply Hong Kong projects and a variety of dredging plant, techniques and contractual arrangements have been employed. Although, as with the Disneyland Resort project, some contractors working on Hong Kong reclamations have used their own dredging plant to source sand from outside Hong Kong, the commonest arrangement has been for outside parties to act simply as suppliers and bring sand into Hong Kong for sale to contractors working on reclamation projects. In terms of volume of imported sand, most was dredged by very small bucket ladder dredgers working sand deposits just south of Hu Men in the Pearl River estuary. This area is where the Pearl River opens out into the estuary proper and where, as a result of the reduced current speed, large volumes of sand bedload were deposited. The transportation of the dredged sand to Hong Kong was mostly by small, self-propelled river barges equipped with a conveyor-belt for discharging over the bows – this lends a distinctive appearance to the vessels from which came their nickname, "pelican barges". These Hu Men sand deposits contain typically less than 5% fines, a composition which, unlike trailer-dredged material, is little changed by the dredging and placing techniques. Lesser volumes of sand have also come from the Chinese mainland side of Mirs Bay which lies to the east of Hong Kong. Here a clean, alluvial sand has at times been extracted by trailer dredger, but mostly by grab-dredgers loading barges for shipment to Hong Kong.

7.6 MUD DREDGING AND DISPOSAL

7.6.1 Contaminated mud

At the start of the 1990s, at an early stage in the planning of how to deal with the more contaminated category of dredged material, it was concluded that disposal in a seabed pit followed by capping with uncontaminated mud would probably provide the best disposal option. There was only one potentially suitable seabed pit already in existence at that time – left as a result of sand extraction for the Tin Shui Wai project mentioned earlier (Figure 7.9). To determine whether this pit would be suitable for the disposal of contaminated mud, disposal experiments were carried out using uncontaminated mud. A large trailer dredger that was employed on uncontaminated mud dredging for the new airport project was used for the trial. The trial involved both simple bottom discharge and placing of the mud down the suction pipe, but measurements indicated that the combination of 20 m water depth and the relatively strong tidal currents resulted in too great a loss of sediment – it will be remembered that these pits were located in one of the natural deep tidal channels. It was therefore determined that an area of shallower water with relatively weak tidal currents was required. Also, because most of Hong Kong's storm weather is associated with intense tropical cyclones or typhoons, coming generally from a southeasterly to southerly direction, shelter from these intense storms was also desirable. The shallow area finally selected in 1992 and still in use today, East Sha Chau, is located not far from the new airport (Figure 7.9). It is a sheltered area with relatively low tidal currents and only 5 to 6 metres of water, but because there was no pre-existing pit in this area at that time, a pit had to be specially dredged for the disposal of the contaminated mud (Whiteside et al., 1996). The environmental acceptability of the disposal facility was established through studies to assess the environmental impact and thereafter, actual performance was measured by a special EM&A programme. Initial monitoring of bottom discharge by barges confirmed there was no unacceptable build-up of pollutants in the area and disposal continued. When filling of the pit was complete, further pits were formed and the monitoring programme progressively developed into a very comprehensive exercise including sediment and water quality, aquatic biota and biological effects testing (Shaw et al., 1998). Paradoxically, one of the advantages of the site chosen for disposal is that it is relatively free from pollution. This means that the field monitoring is more readily able to identify potentially harmful accumulation of contaminants than would be the case were the disposal site in an already polluted area.

The methods of filling and capping the contaminated mud pits (CMPs) were designed firstly to minimise the potential for contaminated sediment in an uncapped pit being remobilised by storm-induced bed shear stresses, and secondly to avoid the possibility of erosion of the completed cap of a pit. It was concluded that, if the highest contaminated mud level was 9 m below sea level, the possibility of remobilisation of contaminated sediment was acceptably low, and if the pit cap was at least 2 m thick the risk of complete erosion of a cap was negligible. It was also important to preclude the possibility of burrowing organisms reaching the contaminated mud – such burrows are commonly less than 0.5 m deep. In the first series of pits a 1 m thick layer of sand was also placed before the 2 m mud cap in order to provide a marker horizon and, notionally, to density the contaminated mud to receive the mud cap. This was principally because extensive use of very high resolution Chirp survey techniques (Selby & Foley, 1995) had revealed the presence of accumulated ponds of fluid mud. As successive pits were capped, investigative coring of completed caps eventually led to the conclusion that the sand layer was unnecessary and its use was discontinued. The overall design therefore became: contaminated mud from the base of the pit (usually about −20 m) up to a level of −9 m and then a 3 m capping layer of uncontaminated mud up to a level of −6 m which approximated to the level of the original seabed. Thereafter, because

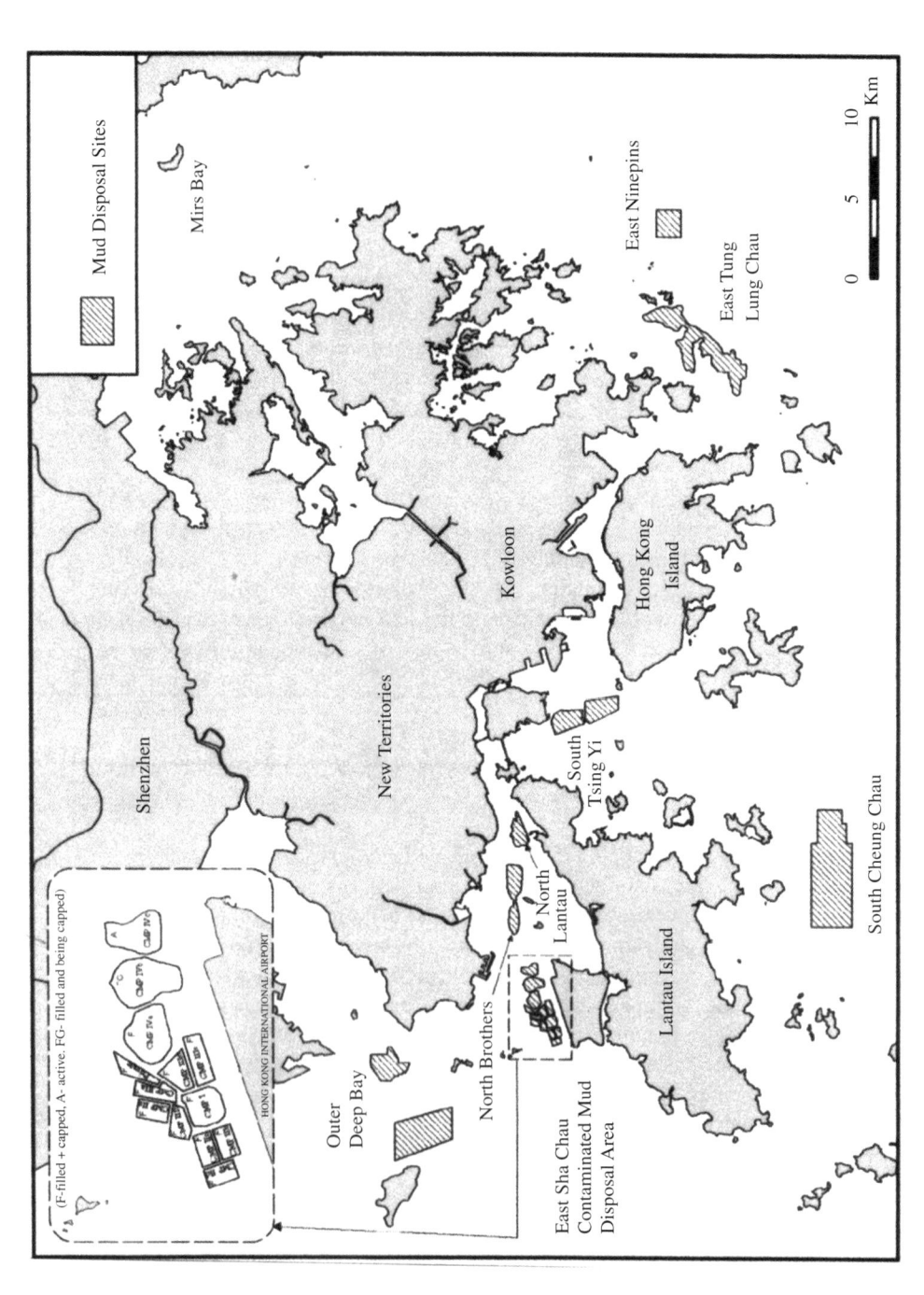

Figure 7.9. Location of mud disposal sites: Contaminated mud pits (East Sha Chau and insert), open seabed disposal mounds (South Cheung Chau & East Ninepins) and backfilling empty sand borrow areas (East Tung Lung Chau, South Tsing Yi, North Lantau, North Brothers & Outer Deep Bay, the latter having been used for Tin Shui Wai New Town project, see earlier text).

of self-weight consolidation of the contaminated mud infill and the capping layer, a final capping layer of 1 to 2 metres thickness is placed to bring the upper surface of the cap back up to the same level as the surrounding seabed. Detailed benthic surveys have since shown that capped pits quickly start to be colonised and that after three years, thriving, diverse communities of taxa similar to adjacent natural seabed are established on the CMP cap (Qian et al., 2003).

The CMPs currently in use were empty seabed pits left after sand extraction for the new airport project. These pits (CMP IVa, IVb & IVc) are larger and deeper than the earlier series of purpose-dredged pits and their suitability and design were covered by a comprehensive EIA in 1997 (ERM-Hong Kong Ltd, 1997).

From the start of the CMP facility, government staff have provided on-site management of the disposal operations, recording incoming vessel details, allocating a specific portion of the disposal site for discharge, and supervising capping works. Figure 7.10 shows the progressive filling of the series of CMPs since late 1992. In total about 40 million cubic metres of dredged contaminated mud has been safely disposed of (equivalent to an *in situ* pre-dredge volume of 30 million cubic metres), and although most of the contaminated mud has been dredged from government projects, private projects with small volumes of contaminated mud needing disposal are also permitted to use the government CMPs. These private projects pay a fee, adjusted from time to time and set to reflect the actual cost of forming, monitoring and capping the pits. To date, the monitoring results have indicated no adverse trends in contaminant levels in any of the monitored pathways and many capped pits are now merging physically and ecologically into the natural background of the seabed. Developing this environmentally acceptable disposal solution compliant with international standards and implementing it without delaying dredging projects has been a major achievement for Hong Kong.

7.6.2 Uncontaminated mud

Mud dredging has taken place to remove overburden at sand borrow areas, to remove soft material for reclamation foundations, and to deepen navigation channels and anchorages. In many cases, plumes of suspended sediment around mud dredging sites have not given rise to the same level of concern as those around sand dredging areas partly because the dark grey sediment plumes were not as visually intrusive as the cream-coloured plumes associated with sand dredging. However, in areas such as reclamation sites where the water depth is not great and where more of the sediment mobilised at the seabed during dredging reaches the surface waters, environmental concerns have been significant. In such shallow areas, it is also not uncommon for propeller wash from trailer dredgers and from tugs pulling barges to contribute very significantly to elevated levels of suspended sediment. In recent projects, considerable efforts have been needed on site to keep levels as low as possible. One general, anti-pollution measure used in Hong Kong is the prohibition of overflow during mud dredging by trailer dredgers and the restriction of the lean mixture overboard system to times when the draghead is off the seabed – such as during manoeuvring.

Three open seabed sites for disposal of dredged mud had been in existence for many years prior to commencement of the major dredging works of the 1990s. One site in eastern waters (Mirs Bay) was abandoned before the major dredging works of the 1990s because of its distant location and because it is situated in an area of important unspoiled natural

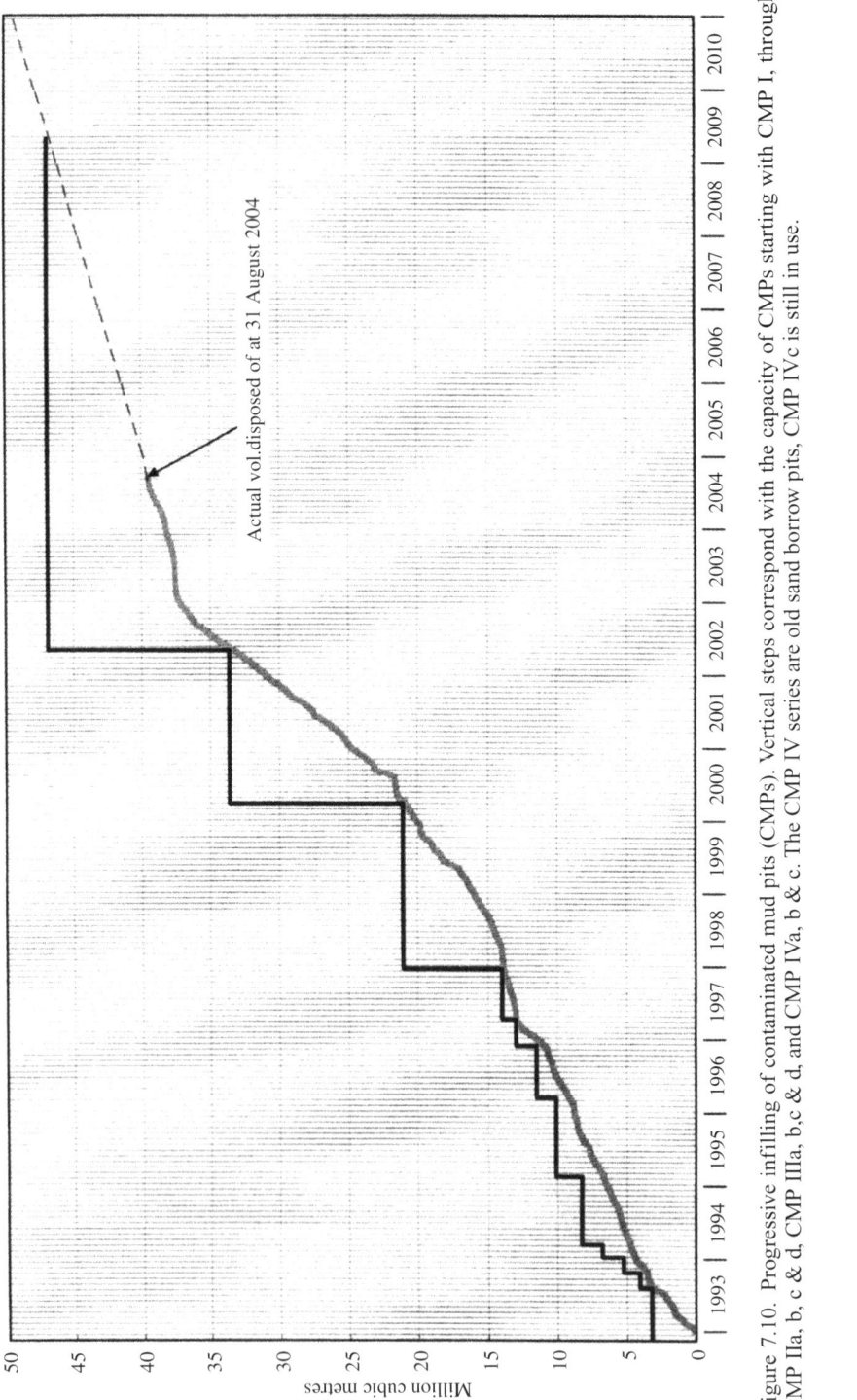

Figure 7.10. Progressive infilling of contaminated mud pits (CMPs). Vertical steps correspond with the capacity of CMPs starting with CMP I, through CMP IIa, b, c & d, CMP IIIa, b,c & d, and CMP IVa, b & c. The CMP IV series are old sand borrow pits, CMP IVc is still in use.

coast. The area was extensively surveyed after being abandoned and taxonomic analysis of very many grab samples illustrated that two years after cessation of disposal operations the site had been colonised by a diverse community of benthic organisms (Valente et al., 1999). The other two disposal sites, shown in Figure 7.9, are still in use and are strategically located in southern waters (South Cheung Chau) and in eastern waters (East Ninepins). These sites were established before EIAs were required and were operated essentially on the basis of maintaining sufficient water depth over the disposal mounds so as not to interfere with vessel movements. By the late 1980s, in parallel with the development of the classification scheme for dredged material, environmental concerns dominated the management of these two sites. Soon all disposal vessels were required to carry a sealed "black box" which recorded the exact position where the disposal took place. Later, water quality monitoring and benthic surveying were undertaken to examine the impacts in the water column and on the seabed. Bathymetric surveys became more frequent and this in turn enabled studies to be undertaken of the mechanisms of consolidation, erosion and spreading which were taking place at the disposal mounds. The sites had not been selected specifically as retention or as dispersion sites, but the studies of the mounds showed a combination of the two characteristics. Mud which had been grab-dredged and placed by bottom-dumping barges retained much of its original strength and tended to remain more or less where placed. Trailer-dredged material on the other hand was much more mobile and weaker. Large areas of slowly flowing mud slurry surrounded parts of the disposal sites, and where continuous disposal in the same location had resulted in the formation of peaks of material, submarine landslides were not uncommon. In one instance (Ng & Chiu, 2001) the sequential bathymetric surveys at the East Ninepins mud disposal site showed that some 420,000 cubic metres of mud had dispersed in a submarine landslide that extended over a kilometre. Initiation of this could be traced to the destabilising effect of the swell associated with a tropical cyclone. In another case, sequential bathymetry showed that a mound of more than one million cubic metres had been completely dispersed by the passing of a severe cyclone. Although these figures are large, they translate into relatively small levels of sedimentation and suspended sediment when viewed in the greater context of the surrounding marine area. And indeed, the monitoring of the disposal sites has not, to date, identified any significant adverse impacts. Nevertheless, the creation of disposal mounds on the seabed was not regarded as ideal and the programme was developed to dispose of mud in empty sand borrow pits when suitable ones become available. Figure 7.9 also shows the locations of exhausted sand borrow pits which have either already been used for mud disposal or which are scheduled for use. It has only been possible for a few of the recent contracts to undertake such backfilling because most contracts completed mud disposal before the borrow areas were exhausted. Nevertheless, as a long-term strategy, the backfilling is continuing with further borrow areas now available for mud disposal. Another reason for backfilling some of the deep sand pits is as a marine safety measure so as to reinstate emergency anchoring capacity for ships – a concern being that moored ships dragging their anchors during severe tropical cyclones might break free if the anchor was dragged into a deep depression. So important a matter was this that a special study involving anchor pulling trials in backfilled mud (Wong & Thorley, 1992) was undertaken to establish the holding ability of mud infill.

7.6.3 Dredging problems caused by unexploded ordnance

During the 1990s, it was not particularly unusual for trailer dredgers in Hong Kong to bring cannon balls, musket balls and in one case, a complete cannon, to the surface. But although these interesting relics of a byegone era can cause problems for dredging plant, they do not give rise to the potentially disastrous consequences associated with their modern, explosive counterparts. Dredging works in Hong Kong have always had the occasional problem of unexploded ordnance dating from the Second World War. These occurrences have been dealt with on an ad hoc basis by the Explosive Ordnance Disposal (EOD) unit of the Hong Kong Police, however, the enormous increase in volumes of dredging from the late 1980s to the mid-1990s resulted in a corresponding sharp increase in the numbers of incidents. Types of ordnance encountered included bombs, mines, ammunition and others. Calls to the Police EOD unit came regularly, often several times per week. Commonest occurrences were the discovery of obvious ordnance or suspicious items when the dragheads of trailer dredgers were brought back on board. The fitting of draghead grills to prevent entry of ordnance into the dredger's pumproom and hopper was standard practice but did not provide a 100% guarantee to exclude all items. Special guidelines and briefings were arranged by the EOD for dredging contractors and their crew as well as improved communications between the two. Although some old charts indicated areas where wartime mines had been laid, little guidance could be given as to locations where such ordnance might now be found, if indeed it still existed. The possible locations of bombs and other ordnance was even more problematic although the concentration of ordnance was clearly greatest in the general harbour area. The most serious incident occurred when the draghead of a trailer dredger hit a large bomb which exploded with sufficient force to dislocate the vessel's main drive gear – a major problem which subsequently contributed to the decision to scrap the vessel. Perhaps because the incident occurred during the night fewer personnel were around and the crew survived the incident.

7.7 FUTURE DREDGING IN THE HONG KONG AREA

Earlier in this chapter, the environmental concerns associated with the actual dredging operations were discussed. There were, however, other public concerns about the major reclamations formed and being planned in the 1990s and these concerns related to the changes in the coastline: islands were being lost by reclaiming between them and most importantly, Hong Kong's famous harbour was becoming ever smaller by continually encroaching reclamation. Such concerns had actually existed in Hong Kong for many decades and Hudson (1979) quotes the 1903 Hong Kong Harbour Master's Report as bemoaning the reduction in the size of the harbour – albeit from a navigation perspective. But recently, the issue came to a head and 1997 saw the enactment of the "Protection of the Harbour Ordinance" which has a presumption against reclamation within the central harbour area. The anti-reclamation lobby has since become stronger and, at the time of writing, several reclamation projects in central Hong Kong are on hold pending further reviews and it seems likely that at least some proposals or parts of proposals may be abandoned.

Although the question of whether or not the central harbour can support more reclamation is important, Hong Kong's broader future lies beyond this area. In a very direct way, Hong Kong's continuing development has been closely linked to that of the adjacent Chinese municipalities even before the 1997 reunification and it is in the context of the tremendous expansion all around the Pearl River estuary and indeed, southern Guangdong Province, that Hong Kong's future development must be viewed (see Figure 7.1).

An early expansion of Hong Kong's port interests was the construction of Yantian container port on the mainland China coast immediately to the east of Hong Kong (see Figure 7.1). The port was constructed by one of the firms operating container terminals in Hong Kong and is already undergoing further expansion by reclamation. But it is to the west of Hong Kong, around the Pearl River estuary, that the greatest future potential lies.

The most notable recent example of a major reclamation project in the Pearl River estuary was the construction of Macau International Airport which between 1992 and 1994, required some 19 million cubic metres of mud dredging and about 35 million cubic metres of dredged sand placement (Carter, 1996). Most of the mud was removed by cutter suction dredgers and interestingly, some of the mud was used to form a reclamation nearby. Sand was brought in by large trailer dredgers working the deeper waters of the outer estuary, by small trailer dredgers working shallower waters upstream, and by small barges loaded by grab-dredgers further up-river.

The most striking future development in the Pearl River estuary is a major bridge, now under detailed planning, from Hong Kong across the estuary to Zhuhai and Macau. Although some dredging will be involved in the bridge construction, especially if an immersed tube is used for the central section, the main dredging works in the area are likely to be related to reclamations and port facilities. On the west side of the estuary, reclamation for port and industry is underway at Nansha. Sand for the reclamation is apparently coming from a nearby site where bucket ladder dredgers are deepening the main navigation channel up the Pearl River. Unlike most reclamations recently constructed in Hong Kong, the soft seabed material has not all been removed – vacuum treatment techniques are being used to accelerate consolidation – and so mud dredging, mud disposal and sand filling have been significantly reduced. On the east side of the estuary, immediately adjacent to Hong Kong, a major new area is to be developed, possibly for port use, at Dachan Bay and this will involve much dredging and reclamation. Shipping access to this area and to the expanding port facilities at Shekou and Chiwan which border Hong Kong will be improved by the dredging of a completely new main navigation channel, the so-called, Tonggu Waterway immediately to the west of Hong Kong's Lantau Island. Elsewhere along the margin of the estuary, major dredging and reclamation is expected for infrastructure, for expansion at Shenzhen Airport, and according to a current Hong Kong study, a possible new container port at the northwestern margin of Hong Kong where a large reclamation or perhaps an artificial island could be required.

From a dredging perspective, there are two key factors in all these major on-going and planned developments around the Pearl River estuary. First, the water depth is generally less than 10 metres – and in many cases, significantly less – so the dredging plant which will be required is going to be rather different from that employed during Hong Kong's heyday of major dredging projects. Second, whether viewed from an environmental or simply cost perspective, it is clearly important to make use of opportunities to combine the need to dredge in one location with the need to deposit material in another. As mentioned above, this is apparently already happening as sand which is dredged to deepen the main Pearl River navigation channel is being used to form a reclamation. In the years to come, large volumes of muddy material will also require to be removed from various locations, perhaps offering scope for formation of long-term reclamations or artificial islands as wetland nature reserves. Whatever unfolds in this thriving area of China, dredging will be a central feature of future developments and the expertise and imagination of dredging engineers will continue to play a key role.

ACKNOWLEDGEMENTS

Past colleagues in the Civil Engineering & Development Department of the HKSAR Government are gratefully acknowledged for providing updates of some key figures. While the data and information used in compiling this account of dredging in Hong Kong is in the public domain, any views expressed are the author's own and do not in any way represent the views of the HKSAR Government.

REFERENCES

Arthurton, R.S. (1987). Studies of Quaternary geology and the exploration for offshore sources of fill in Hong Kong. *Proceedings of the Symposium on the Role of Geology in Urban Development in Southeast Asia (Landplan III), Geological Society of HongKong, Bulletin* no. 3, pp 229–238.
* Binnie Consultants Limited. (1991–3). Fill Management Study – Phase II: Investigation and Development of Marine Borrow Areas. *Borrow Area Assessment Report Series.* Work on nine separate marine borrow areas undertaken on behalf of the Geotechnical Engineering Office, Government of the Hong Kong Government Special Administrative Region.
* Binnie Consultants Limited. (1992). *South Mirs Bay Environmental Impact Assessment – Initial Assessment Report.* Report to the Geotechnical Engineering Office, Hong Kong, 184p.
* Binnie Consultants Limited. (1993). *East Lamma Channel borrow area – Scoped environmental assessment*, Final Report to the Geotechnical Engineering Office, Hong Kong, 131p.
* Binnie Consultants Limited. (1995a). *Hypoxia and Mass Mortality Event In Mirs Bay* Final Report. Report to the Geotechnical Engineering Office, Hong Kong.
* Binnie Consultants Limited. (1995b). *Marine ecology of Hong Kong: Report on underwater dive surveys (October 1991 – November 1994).* Report to the Geotechnical Engineering Office, Hong Kong, 2 vols, 161p.
Brand, E.W. & Whiteside, P.G.D. (1990). Hong Kong's fill resources for the 1990s. *The Hong Kong Quarrying Industry, 1990–2000*, edited by P. Fowler & Q.G. Earle, pp 101-112. Institute of Quarrying, Hong Kong.
Brand, E.W., Massey, J.B. & Whiteside, P.G.D. (1994). Environmental aspects of sand dredging and mud disposal in Hong Kong. *Proceedings of the First International Congress on Environmental Geotechnics*, Edmonton, Alberta, Canada, pp 1–10.
Carter, M. (1996). Methods used to reduce settlements and shorten construction time at Macau International Airport. *The Hong Kong Institution of Engineers Transactions* Vol 3, No 1.
Chalmers, M.L. (1983). Preliminary assessment of sedimentation in Victoria Harbour, Hong Kong. *Proceedings of the Meeting on Geology of Surficial Deposits in Hong Kong*, pp 117–129. (Published as *Geological Society of Hong Kong Bulletin No. 1*, 1984).
Cheung, S.P.Y. & Ho, J.L.P. (2004). Plume monitoring of dredging at marine borrow area in Hong Kong. *Proceedings of the International Conference on Coastal Infrastructure Development*, Hong Kong, 22–24 November 2004 (In press).

de Kok, M.B.A.M. (1994). Hong Kong's New Container Terminal Eight. *Terra et Aqua 57*, pp 26–29.
* DEMAS, (1996). *Assessment of remaining sand reserves at West Po Toi marine borrow area*. Report to the Geotechnical Engineering Office, Civil Engineering Department of the Government of the Hong Kong Government Special Administrative Region.
* Dredging Research Ltd. (1995). Second Territorial Suspended Sediment Survey (Dry Season) – Draft Report Volumes 1 & 2. Report to the Geotechnical Engineering Office, Civil Engineering Department, Hong Kong, 28p, plus Appendix.
Dutton, C. (1987). Marine fill investigation for site formation at Tin Shui Wai in Hong Kong. *Hong Kong Engineer*, vol 15, No.9, pp 29–38.
+ Environment, Transport and Works Bureau (2002). Technical Circular No. 34/2002 "Management of Dredged/Excavated Sediment", Environment, Transport & Works Bureau, Government of the Hong Kong Special Administrative Region.
* ERM-Hong Kong Ltd. (1997). Environmental impact assessment for disposal of contaminated mud in East Sha Chau marine borrow pit. Report to the Civil Engineering Department, Government of Hong Kong Special Administrative region.
Evans, C.D.R. (1988). Seismostratigraphy of early Holocene sand banks. In *Marine Sand and Gravel resources of Hong Kong, Proceedings of the Seminar on Marine Sources of Sand* (Whiteside & Wragge–Morley, Eds) *Geological Society of Hong Kong*, pp 45–52.
Evans, N.C. (1994). Effects of dredging and dumping on the marine environment of Hong Kong. *Terra et Aqua 57*, pp 15–25.
* EVS Environment Consultants (1996a). Review of Contaminated Mud Disposal Strategy and Status Report on Contaminated Mud Disposal Facility at East Sha Chau. Report to the Geotechnical Engineering Office of the Hong Kong Government.
* EVS Environment Consultants (1996b). Classification of dredged material for marine disposal. Report to the Geotechnical Engineering Office of the Hong Kong Government.
Fyfe, J.A., Shaw, R., Campbell, S.D.G., Lai, K.W. & Kirk, P.A. (2000). *The Quaternary Geology of Hong Kong*, Hong Kong Geological Survey, Geotechnical Engineering Office, Government of the Hong Kong SAR, 209p plus maps.
Germano, J.D., Reid, C.A., Whiteside, P.G.D., & Kennish, R. (2002). Field verification of computer models predicting plume dispersion in Hong Kong. In: S.Garbaciak Jr. (Ed). *Dredging '02. Key Technologies for Global Prosperity*. Proceedings of the third specialty conference on Dredging and Dredged Material Disposal. May 5–8, 2002, Orlando, Florida. Sponsored by: Coasts, Oceans, Ports, and Rivers Institute (COPRI) of the American Society of Civil Engineers (ASCE). ISBN 0-7844-0680-4.
Guilford, C.M. (1988). Marine fill in Hong Kong – a 35 year resume. In *Marine Sand and Gravel resources of Hong Kong, Proceedings of the Seminar on Marine Sources of Sand* (Whiteside & Wragge–Morley, Eds) *Geological Society of Hong Kong*, pp 11–22.
Holmes, P.R. (1988). Environmental implications of exploiting marine sand. In *Marine Sand and Gravel resources of Hong Kong,* Proceedings of the Seminar on Marine Sources of Sand (Whiteside & Wragge–Morley, Eds) *Geological Society of Hong Kong*, pp 143–159.
Hudson, B.J. (1979). Coastal Land Reclamation with Special Reference to Hong Kong. *Reclamation Review* Vol 2, Pergamon Press Ltd, pp 3–16.
IMO, (1995). *Dredged Material Assessment Framework (Resolution LC.52 (18))*. Adopted at the 18th Consultative Meeting of the London Convention, 4–8 December 1995. International Maritime Organisation, London.
Ip, K.L. (1996). "Victoria Harbour, Western Harbour and North Lantau Waters". In *Coastal infrastructure development in Hong Kong, Proceedings of the 1995 Symposium on the Hydraulics of Hong Kong Waters*. Civil Engineering Department, Hong Kong Government, pp 33–65.
* Kwan, S.H. (1993). *Review of specifications for marine fill material for reclamation*, Geotechnical Engineering Office internal report, Technical Report TN3/93, Hong Kong.
Land, J., Kirby, R. & Massey, J.B. (1994). Recent innovations in the combined use of acoustic doppler current profilers and profiling siltmeters for suspended sediment monitoring. *Proceedings of the 4th nearshore and estuarine cohesive sediment transport conference.* Wallingford, UK
Lei, P.C.K., Fok, A.W.K., Dawes, A. & Whiteside, P.G.D. (1998). Development of Hong Kong's decision criteria for sediment disposal, *Proceedings of the Seventh International Symposium on River*

Sedimentation and Second International Symposium on Environmental Hydraulics, University of Hong Kong.

Leung, K.F. & Morton, B. (2000). The 1998 Resurvey of the Subtidal Molluscan Community of the Southeastern waters of Hong Kong, Six Years after Dredging Began and Three Years Since it Ended. In: *The Marine Flora and Fauna of Hong Kong and Southern China*. Ed. B. Morton. Proceedings of the Tenth International Marine Biological Workshop, Hong Kong 6–26th April 1998. Hong Kong University.

Malone, A.W. & Oakervee, D.E. (1993). Providing aggregates for a major development, Hong Kong container port and airport projects. *Quarry Management*, vol 20, November 1993, pp 11–17.

Martin, R.P., Whiteside, P.G.D., Shaw, R. & James, J.W.C. (1997). Offshore Geological Investigations for Port and Airport Developments in Hong Kong. *Proc. 30th Int'l Geol Congr.*, Vol 23, pp 65–80.

Ng, K.C., Whiteside, P.G.D. & Kwan, S.H. (1998). Physical and ecological studies on effects of sand dredging and mud disposal in Hong Kong. In: *Proceedings of the Second International Conference on the Pearl River Estuary in the Surrounding Area of Macau*, Guangzhou and Macau, pp 187–202.

Ng, K.C & Chiu, K.M. (2001). Back-analysis of a large submarine landslide at a marine disposal mound. *Proceedings of the Third International Conference on Soft Soil Engineering* (Lee, C.F., Lau, C.K., Ng, C.W.W. Eds), Hong Kong, 6–8 December 2001, A.A. Balkema Publishers, pp 269–274.

Ng, K.C. & Chan, R.K.S. (2004). Effects of sand dredging and mud disposal on the marine environment in Hong Kong. *Proceedings of the International Conference on Coastal Infrastructure Development*, Hong Kong, 22–24 November 2004 (In press).

Ooms, K., Woods, N.W. & Whiteside, P.G.D. (1994) Marine sand dredging: key to the development of Hong Kong. *Terra et Aqua*, 54, pp 7–16.

Parry, S. (2001). Natural and anthropogenic effects of offshore suspended sediment loads in Hong Kong: Implications for dredging. *Proceedings of the 14th South-East Asian Geotechnical Conference*, Hong Kong, December 2001. Balkema Publishers. pp 395–399.

Plant, G.W., Covil, C.S. & Hughes, R.A. (Eds) (1998). *The site preparation for the new Hong Kong International Airport: The design, construction and performance of the airport platform* Thomas Telford, London.

Qian, P-Y., Qiu, J-W., Kennish, R. & Reid, C.A. (2003). Recolonisation of benthic infauna subsequent to capping of contaminated dredged material in East Sha Chau, Hong Kong. *Estuarine, coastal and shelf science*, 56 (2003), pp 819–831.

Ridley Thomas, W.N., Lai, M.W.C. & Nieuwenhuijs, G.K. (1988). Marine geophysical methods. *Proceedings of the Seminar on Marine Sources of Sand*, Hong Kong, pp 109–120. (Published under the title *Marine Sand and Gravel Resources of Hong Kong*, edited by P.G.D. Whiteside & N. Wragge-Morley, Geological Society of Hong Kong, 1988).

Rodger, J.G., (1996). "Selected tidal hydraulic and marine environmental studies 1980–1995". In: *Coastal infrastructure development in Hong Kong*, Proceedings of the 1995 Symposium on the Hydraulics of Hong Kong Waters. Civil Engineering Department, Hong Kong Government, pp 311–324.

Selby, I & Foley, M. (1995). An application of Chirp acoustic profiling: monitoring dumped muds at seabed disposal sites in Hong Kong. *Journal of Marine Environmental Engineering*, vol 1, pp 247–261.

Selwood, J.R. & Whiteside, P.G.D. (1992). The use of GIS for resource management in Hong Kong. *Proceedings of the 8th Conference on Computing in Civil Engineering*, American Society of Civil Engineers, pp 942–949.

Shaw, J.K., Whiteside, P.G.D. & Ng, K.C. (1998). Contaminated mud in Hong Kong : A case study of contained seabed disposal. XV World Dredging Congress, Las Vegas, pp 799–810.

Shaw, R. (1988). The nature and occurrence of sand deposits in Hong Kong waters. In *Marine Sand and Gravel resources of Hong Kong, Proceedings of the Seminar on Marine Sources of Sand* (Whiteside & Wragge–Morley, Eds) *Geological Society of Hong Kong*, pp 33–43.

Styles, K.A. & Hansen, A. (1989). *Territory of Hong Kong*, Geotechnical Area Studies Programme (GASP) Report XII, Geotechnical Control Office, Hong Kong, 346p.

Valente, R.M., McChesney, S.M. & Hodgson, G. (1999). Benthic recolonisation following cessation of dredged material disposal in Mirs Bay, Hong Kong. *Journal of Marine Environmental Engineering 5*, pp 257–288.

Vlasblom, W.J. (1994). European dredgers build new Hong Kong airport. *Terra et Aqua, 54*, pp 3–6.

Whiteside, P.G.D. (1991). Management of Hong Kong's marine fill resources. *Reclamation – Important Current Issues*, Hong Kong Institution of Engineers, pp 33–47.

Whiteside, P.G.D. & Massey, J.B. (1992). Strategy for exploration of Hong Kong's offshore sand resources. *Proceedings of the International Conference on the Pearl River Estuary in the Surrounding Area of Macao*, Macau, vol 1, pp 273–281.

Whiteside, P.G.D. & Rodger, J.G., (1996). "Application of numerical modelling to sand dredging and mud disposal". In: *Coastal infrastructure development in Hong Kong, Proceedings of the 1995 Symposium on the Hydraulics of Hong Kong Waters*. Civil Engineering Department, Hong Kong Government, pp 361–372.

Whiteside, P.G.D., Ng, K.C.S. & Lee, W.P. (1996). Management of Contaminated Mud in Hong Kong, *Terra et Aqua, 65*, December 1996, pp 10–17.

Whiteside, P.G.D., Ooms, K. & Postma, G.M. (1996). Generation and decay of sediment plumes from sand dredging overflow, *Proceedings of the XIVth World Dredging Congress*, Amsterdam, Central Dredging Association, Delft, The Netherlands, pp 877–892.

Whiteside, P.G.D., Massey, J.B. & Lam, B.M.T. (1998). Marine fill – the key to Hong Kong's airport Core Projects, in *Geotechnical Aspects of the Airport Core Projects*, Hong Kong Institution of Engineers, pp 97–108.

Whiteside, P.G.D., Ding, G.W.W., Lei, P. & Tsing, M.M.K. (2001). New classification scheme for dredged material disposal in Hong Kong. *Proceedings of the XVIth World Dredging Congress*, held in Kuala Lumpur 2–5 April 2001.

* Wong, C.K. & Thorley, C.B.B. (1992). *Backfilled Mud Anchor Trials Feasibility Study*. Geotechnical Engineering Office, Hong Kong, 68p. (GEO Report No. 18).

Wragge–Morley, N. (1988). Dredging for Container Terminal 6: A case history. In *Marine Sand and Gravel resources of Hong Kong, Proceedings of the Seminar on Marine Sources of Sand* (Whiteside & Wragge–Morley, Eds) *Geological Society of Hong Kong*, pp 121–129.

* Reports which can be consulted in the Civil Engineering Library of the Civil Engineering & Development Department of the Government of the HKSAR

+ At the time of writing, available within the HKSAR Government website. See index page at http://www.info.gov.hk/eindex.htm

Chapter 8

Singapore Dredging

Weidong Sun
HR Wallingford, Wallingford, UK

8.1 INTRODUCTION

The Republic of Singapore is a focal point for trade and economic development in the region, with one of the world's busiest ports and largest oil refineries. The income generated from industry and shipping traffic has helped it to become one of the wealthiest nations in the region. The population grew rapidly from 200 in 1819 to more than 3 million today, in a country with a land area of only 583 km^2, inclusive of its 57 offshore islands in its dependence in 1965 (Figure 8.1).

The resulting high population density and fast industrial development have created a constant need to reclaim land from the foreshore for various development projects along the coastal line and various other dredging works. This has resulted in some of the biggest dredging and reclamation projects the world has ever seen.

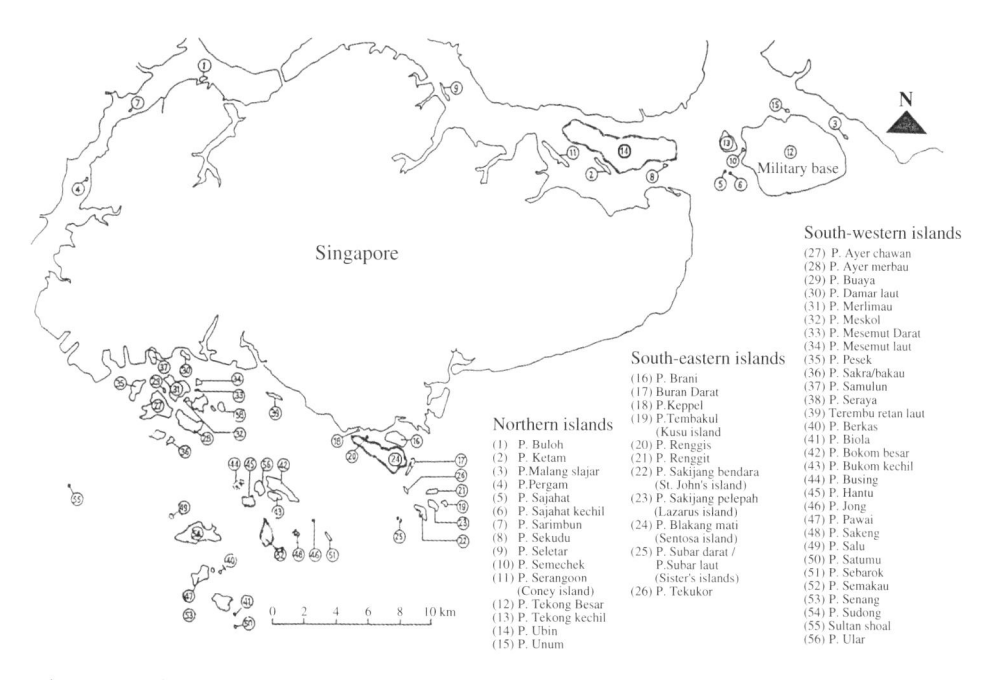

Figure 8.1. Singapore and it's offshore islands in 1959.

8.1.1 Meteorological, hydro-graphic and geological conditions

The tides in Singapore waters are of mixed type, they are predominantly semi-diurnal with mean and maximum tidal ranges of 2.3 m and 3.4 m respectively. The currents, which are governed by the tides, are generally weak with the speed ranging from 0.1 m/s to 0.2 m/s nearshore and with the maximum speed of about 2 m/s in deep water. Wind conditions are governed by two monsoon seasons. During the NE monsoon season from December through March, the prevailing wind directions are N and NNE. During the SW monsoon season from July through October, the prevailing wind directions are S, SSE and SE. Wind speeds only very rarely exceed 10 m/s. The waves off the southeast coast consist partly of wind generated waves and partly of refracted swell penetrating from the east from the South China Sea during the NE-monsoon period. The maximum significant wave height is 0.6 m.

In recent years, the net longshore sediment transport rate is low, estimated at 2 tonnes/day along the east coast in the mid to late 1990's (Shankar and Sun, awaiting publication, 2004) and even less today. Elsewhere in Singapore, longshore drift is negligible. Nowadays, coastal sediments are produced mainly by dredging activities, especially land reclamation.

8.2 DREDGING ACTIVITIES

Dredging activities in Singapore consist of land reclamation, port development, regular maintenance of rivers, storm drains and navigation channels, beach nourishment, pipeline laying and sand mining. Where the material is not dredged for a specific use it has to be disposed of in an appropriate manner.

8.2.1 Land reclamation

The resulting high population density has created a constant need to reclaim land from the foreshore for various development projects along the coastline, especially for airports, seaports, maritime and petroleum industries, housing, commerce and recreation parks (Figure 8.2).

The massive land reclamation started in 1965 and has not stopped since. Singapore has increased its size by 18%, from 583 km^2 in 1965 to more than 680 km^2 in 2003 (Figure 8.3), and average of 7,000 m^2/day. The calm wave environment and rather low tidal range prevailing along the coastline of Singapore, has enabled the reclamation work to be executed quite easily and economically.

8.2.2 Maintenance dredging of navigation channels and ports

Regular maintenance dredging of navigation channels, ports and their approaches were one of main dredging activities in Singapore waters for many years. However, in recent years, the nearshore seabed has remained virtually unchanged due to very low sediment transport rates and maintenance dredging is hardly necessary.

8.2.3 Dredging for beach creation/nourishment

Sandy beaches were built during 1966 and 1971 on the southeast coast, mainly for housing and recreation. The beaches are protected with headland breakwaters. Beach nourishment

SINGAPORE DREDGING 203

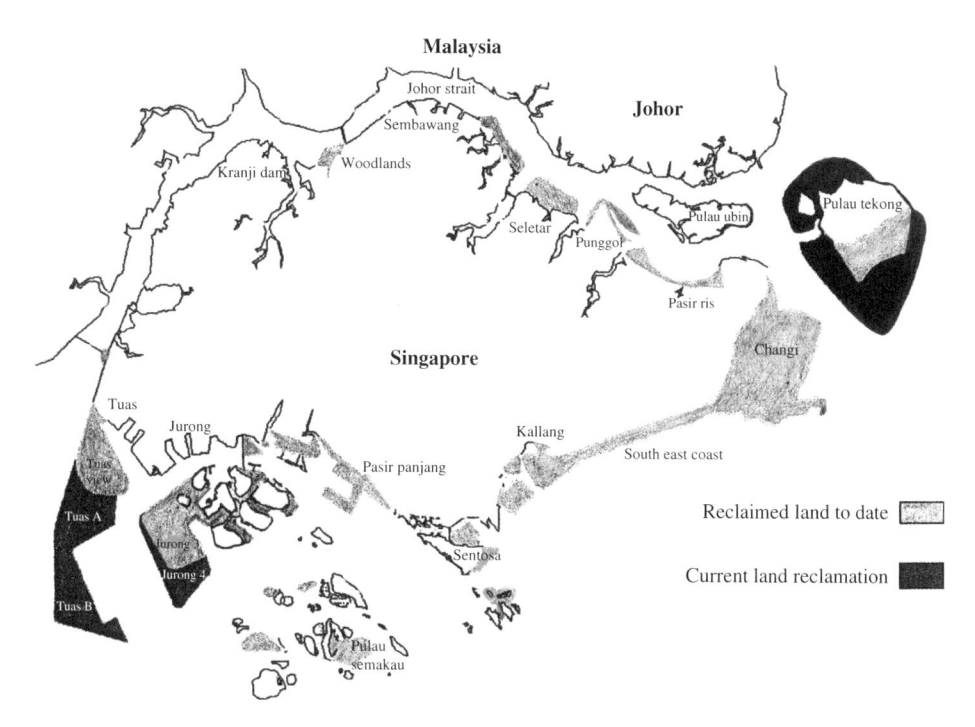

Figure 8.2. Land reclamation in Singapore in December 2003.

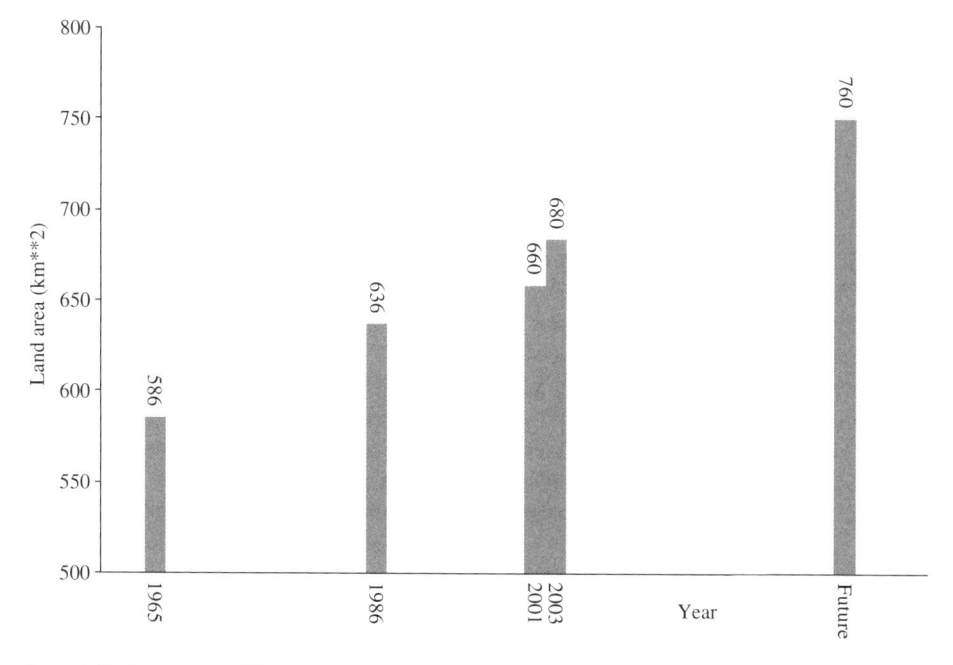

Figure 8.3. Land area of Singapore.

took place at, for example, Bedok in the 1980s. The beach has remained in equilibrium throughout the 1990s (Shankar and Sun, awaiting publication, 2004).

By March 1997, two breakwaters were built at Siloco beach, on the southwest shore of Sentosa Island, at the southern end of Singapore, to protect the sandy beach from erosion by the wash from the high-speed ferries plying between the nearby World Trade Centre and the Riau islands of Indonesia. Breakwaters are needed only at the above two beaches; elsewhere in Singapore, longshore drift is negligible and beach nourishment has not been needed to date.

8.2.4 Dredging from outfalls of storm drain

Rainwater flows to the sea through storm drains all over Singapore. Sand accretion at the mouths of the storm drains decrease the capacity of discharge which sometimes results in flooding in city areas (Chen and Sun, 1999). Regular maintenance dredging is carried out by small dredgers taking several weeks to several months depending on location and quantity. However, the amount of sand dredged is very small.

8.2.5 Sand mining

In the 100 years up to 1986, nearly 165 million m^3 of earth from the land and 120 million m^3 of sand from the seabed were moved to reclaim 53 km^2 of land (Pui, 1986). Singapore will need 1.8 billion m^3 of seabed fill between 2001 and 2010 for land reclamation and construction industry (Moch and Tiarma, 2002). The main sources of material were earth obtained by levelling hills, seabed sand, marine clay, sand arising from maintenance dredging of navigation channels and recycled construction material.

8.2.5.1 *Material by levelling hills*
Up to 1974 the material used for land reclamation was obtained entirely from levelling hills, an operation carried out in phase with the development of public housing and industrial estates. The borrow area used for the East Coast Phases III and IV Reclamation projects was subsequently converted into a fresh water reservoir (Wei and Chiew, 1983).

8.2.5.2 *Seabed sand*
As land based sources were rapidly diminishing in capacity, seabed sand was more in demand for land reclamation. For the reclamation of islands, seabed sand was the natural fill material. Dredging of the seabed has beneficially enabled ships with deeper draught to call at Singapore port, the world's busiest port.

As the availability of seabed sand in Singapore waters became depleted sources and concessions were sought from the neighbouring countries. The distances to the borrow sites range from 40 km (Pulau Batam and Pulau Bintan, Indonesia) to 400 km (Eastern Sumatra and Bangka and Belitung, Indonesia) (Figure 8.4).

Seabed sand was also dredged as aggregates for the Singapore construction industry.

8.2.5.3 *Recycled material from construction sites*
The sand used to reclaim land is imported and costly. Construction waste was often used for fill material. For example, about 80% (or 7.6 million m^3) of soil excavated from the building of the $5 billion Mass Rapid Transport line on the land of Singapore was used for embankments and back-filling.

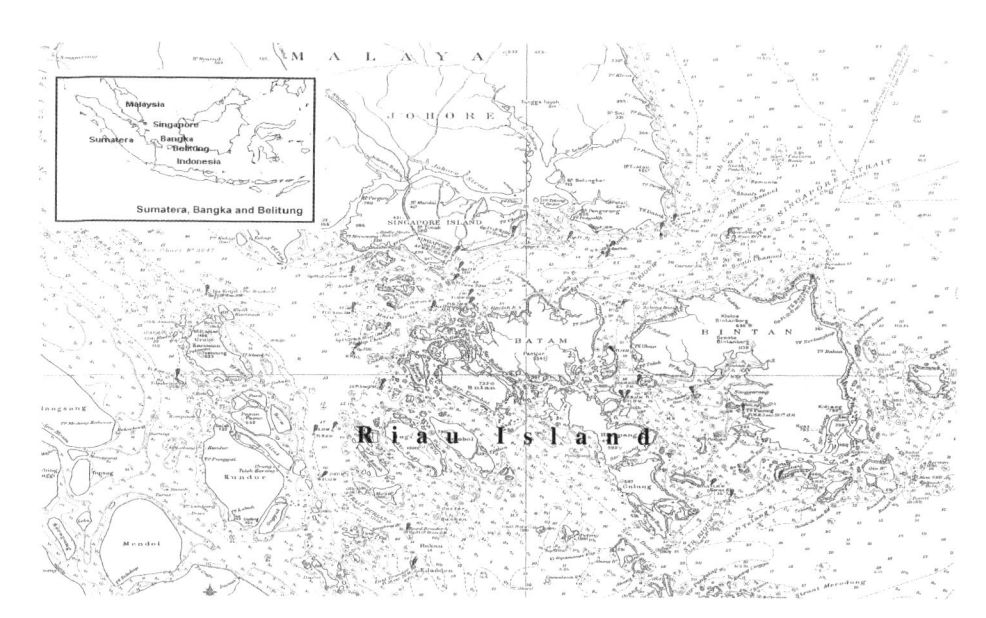

Figure 8.4. Sandy borrowing site.

8.2.5.4 *Marine clay*

Singapore waters have rich marine clay resources. Clay was previously not used extensively in reclamation because it takes many years to dry and consolidate. A study was conducted in which thin sand layers were interspersed with the marine clay. The result showed that settlement of the fill could be accelerated and the strength improved (Lee, 1987).

8.2.6   Disposal of dredged material

Dredging is frequently carried out in Singapore for routine maintenance of its navigation channels and for port construction activities. The disposal of large quantities of dredged material poses a severe problem in Singapore as there are limited approved disposal sites (Leung et al, 2001).

Part of the material arising from construction was used together with marine clay as fill material for land reclamation. The remainder has been placed offshore of Singapore and its neighbouring countries.

From May 1995 to 1999, the Ministry of Environment worked on a 7 km long perimeter bund and wharf at Pulau Semakau. This created a landfill area about 350 hectares which could contain about 63 million m^3 of refuse and dredged material.

8.3   DREDGING PROJECTS

8.3.1   Earlier reclamation works

Early foreshore reclamation/dredging works were mainly confined to the southern tip of the main land of Singapore, from Kallang Basin at the east to Tanjong Berlayer at the West.

The total amount of land area reclaimed before self-rule in 1959 was about 3 km^2 (Pui, 1986).

The first major foreshore reclamation works using dredging was in what was then Telok Ayer Bay about 100 years ago, for the construction of a link road between the commercial centre at Singapore River and a new deep water port at Keppel Channel. Fill material was obtained by levelling 2 hills at Tanjong Pagar (Bogaars, 1956). Covering approximately 7 ha, the project took more than 8 years to complete.

This reclaimed land was subsequently further extended and completed in 1915 with the formation of Telok Basin.

Early reclamation works also included reclaiming 9 ha of land at the present Container Terminal by the Tanjong Papar Land Company in the 1880's, and reclamation along Keppel Channel in the 1900's for the construction of a seaport and shipyards.

At the confluence of the Rivers Kallang and Geylang, about 1 km^2 of land was reclaimed in the 1930's for the construction of Kallang Airport. 7 million m^3 of fill material was obtained from a hill situated some 6 km away from the site. Trains were used for the transportation of the fill material. This project took 4 years to complete. During the same period, the coastline from the mouth of Singapore River to the mouth of Rochore River was also reclaimed for the construction of an esplanade.

8.3.2 Reclamation works since 1959

With the achievement of self-rule in 1959, and subsequent independence in 1965, various reclamation works were initiated to cater for rapid development in public housing, industry, commerce, recreation, airport and seaport development, and for construction of express ways (Pui, 1986).

Land reclamation has modified the coastline of Singapore, extending it seawards, especially on the eastern, north-eastern and western parts of the island, and changing it quite beyond recognition. Large coastal areas have also been straightened by building dykes across estuaries, particularly on the west coast across the estuaries of Tengeh, Poyan, Murai and Sarimbun. Many offshore islands have become larger.

Reclamation works are generally carried out by three Statutory Boards as agents to the government, namely Jurong Town Corporation (JTC), Housing and Development Board (HDB) and the Port of Singapore Authority (PSA). Together, by 1986, these agencies had reclaimed a total about 48 km^2, representing more than 90% of the total land area reclaimed from foreshore during the 100 years up to that time. From 1965 to 1986, Singapore grew in size at a rate of 8,000 m^2 per day (Radhakrishnan et al, 1983).

8.3.3 JTC land reclamation projects

Land reclamation works conducted by JTC, mainly for industrial development, are Tuas foreshore reclamation, Jurong Port extension, Loyand Reclamation, Senoko Reclamation, reclamation at Seraya Island, Ayer Merbau Island, Busing Island, Sakra/Bakau Islands and currently the world's biggest land reclamation project Jurong Island Reclamation and Tuas View Extension. Brief information is given about these in the following paragraphs.

8.3.3.1 *Tuas View Reclamation*
From 1983 to 1988, JTC spent US$2.1 billion in reclaiming 600 hectares of land at the offshore area of Tuas at the western end of Singapore (Figures 8.2 and 8.5). The swamps in the Sungai Buloh area were transformed into an industrial estate.

8.3.3.2 *Jurong Island Reclamation*
The seven islands that used to exist off the southwest coast of Singapore, Pulau Merlimau, Pulau Ayer Chawan, Pulau Ayer Merbau, Pulau Seraya, Pulau Pesek and Pulau Pesek Lecil are being combined into a single island, Jurong Island. This amalgamation is being carried out by phased reclamation of the channels between the islands (which had water depths ranging from 0–20 m ACD) and extending into the sea to form one big island. When completed, Jurong Island will form a land area of about 3,200 hectares from an initial area of less than 1,000 hectares and will represent approximately 4.7% of the land area of Singapore. The entire project is expected to cost US$5 billion (US$2.8). The sand used was mined from offshore of Riau Province, Indonesia, Malaysia and the South China Sea.

8.3.3.3 *Tuas View Extension*
Straight across from Jurong, the Tuas View Extension A (975 hectares, 260 million m^3 fill) and Extension B (933 hectares, 395 million m^3) are under construction (Figure 8.5).

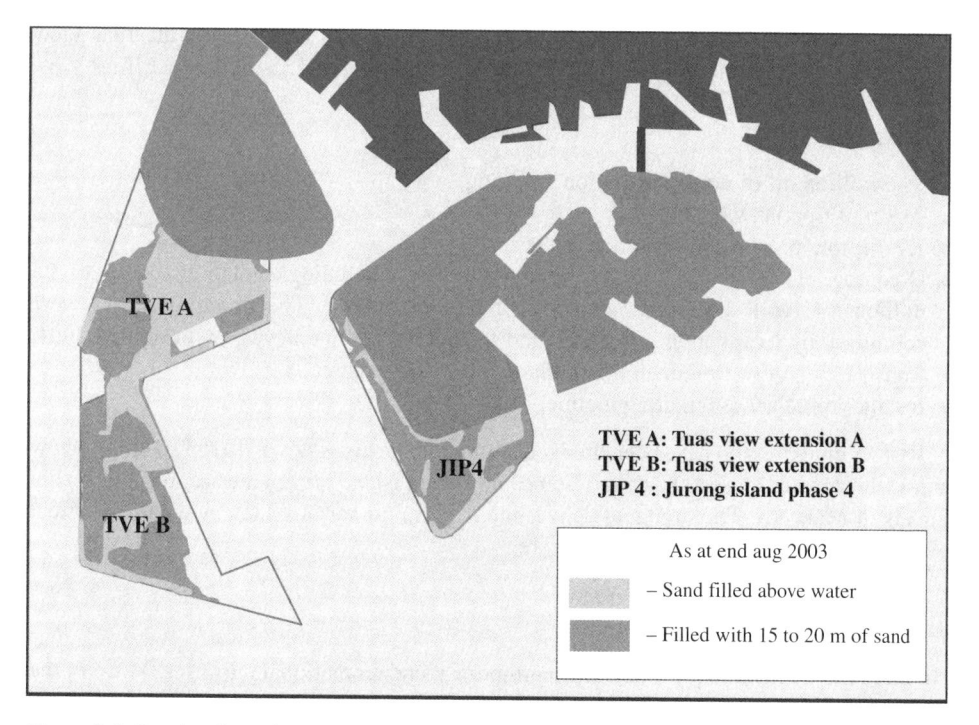

Figure 8.5. Land reclamation at Jurong Island and Tuas View.

Summary data: phases of reclamation of Jurong Island and Tuas View.

Phase	Area reclaimed (ha)	Contractor(s)	Cost US$ millions	Features
1 1991–1998	218	Penta Ocean Koon Construction Dredging International	150	Height 5 m ACD Side Slopes 1:3–1:4 Included a causeway linking the island to the Singapore mainland.
2	372	Dredging International Penta Ocean Koon	550	
3 1998–2002	1000	Penda, Koon, Ham, Dredging International and BKI	800	218 million m^3 included construction of about 11 km sandkey.
4 2000–2005 est.	547 off Jurong 917 in Tuas View	Penta Ocean Koon Construction Boskalis and Ham	2,100	550 million m^3 Largest job in the history of dredging. Area to be reclaimed represents 2.2% of the size of Singapore.

This last phase, still under construction and due to finish in 2005 represents the biggest dredging contract ever awarded. It consists of reclaiming 547 hectares off Jurong Island and building a large chemical island, as well as reclaiming 917 hectares in the Tuas View area, giving a total earth volume of 550 million m^3 (Jurong 4: 155 million m^3, Tuas View Extension: 395 million m^3).

The project also includes:

- 55.5 million m^3 of non-reclamation dredging,
- 24 km of riparian works,
- 1.7 million m^3 of stone revetment works,
- Dredging of waterways around Jurong 4 project including capital dredging of 25 million m^3, with a subcontract value of US$49 million. The dredging works were completed by Joint venture: Boskalis and Jan De Nul JV in one year by the end of 2001.
- Construction of two piers on the south of the reclaimed area,
- Installation of a 7,500 metre pipeline.

Vast volumes of sand have been dredged, brought from distances of 40 to 110 km. Borrow areas were located close to Karimun Island and Batam Island off Indonesia, and at Ramunia Shoals in Malaysia. The dredgers deliver the material to various sites by means of direct placement, rainbowing or pumping ashore.

8.3.4 HDB land reclamation projects

HDB reclaim land mainly for housing, commercial and recreational purposes. Much of the information in the following list was obtained from the Housing Development Board web site (HDB, 2004).

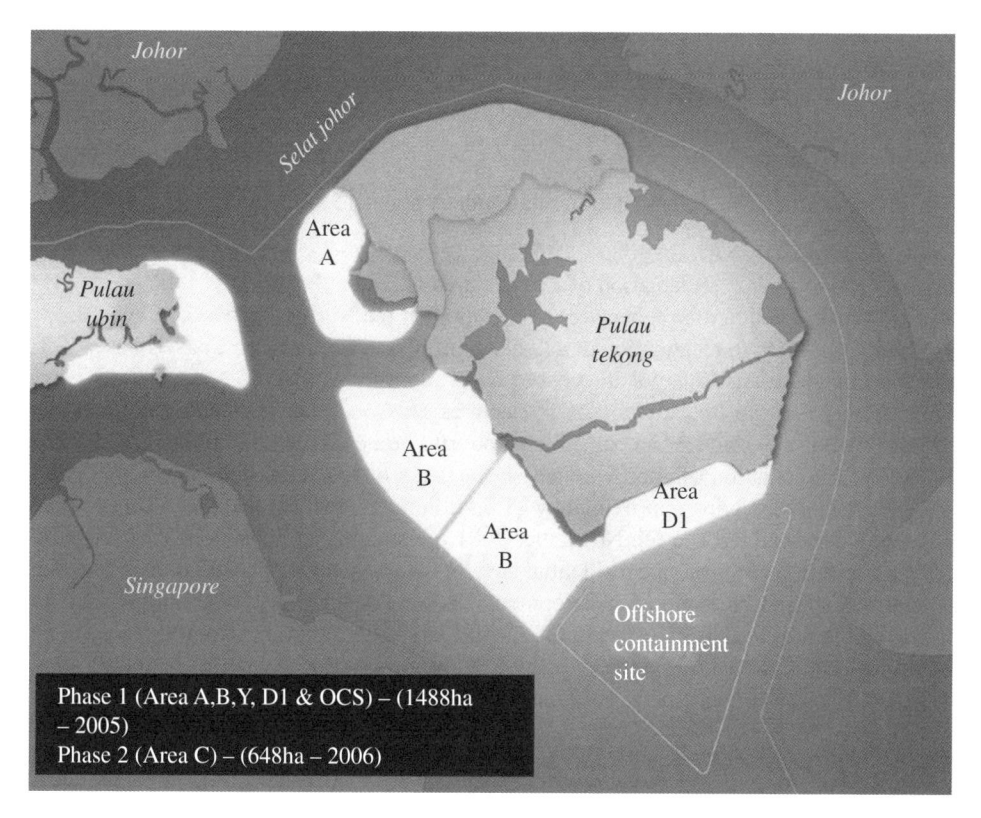

Figure 8.6. Land reclamation at Pulau Tekong and Pulau Ubin.

8.3.4.1 *South-east coast*

From 1965 to 1971, the HDB reclaimed the entire south-eastern coastline of Singapore main island, from Changi at the east to the city centre at the south; which added a total of 1,525 hectares (Pui, 1986) (Figure 8.2). These works had been executed in 7 major phases, in phase with public housing development schemes during which hills at Bedok and Tampines were levelled and the material used for the building of new towns.

8.3.4.2 *Pulau Ubin and Pulau Tekong*

The HDB reclaimed land around Pulau Ubin and Pulau Tekong for long-term urban development needs. The reclamation will be carried out in three separate phases at an estimated cost exceeding US$1.7 billion. Work commenced in January 2000 and the whole project was targeted to be completed by December 2008. The final reclamation area will be 3,306 hectares (Figures 8.2 and 8.6).

Phase 1 of this is in progress at the time of writing (2004). This is to reclaim 1,218 hectares for the expansion of housing facilities. The US$1.1 billion project started on 9th November 2000 and is expected to be finished by November 2005. The project includes:

- Borrowing 190 million m^3 of sand from offshore of neighbouring countries;
- Dredging 17 million m^3 of material;
- Dredging of waterways about 3.0 million m^3 of capital dredging;

- Constructing an offshore disposal site for about 24 million m^3 of soft clays;
- Placing soft dredged material at reclamation areas;
- Filling with sand (this requires special techniques allowing the sand to be placed carefully without causing disturbance of the underlying layers);
- Construction of a staging ground at the mainland: 6 million m^3 of good earth, 3.5 million m^3 of clay and construction of a sandkey about 15 km long.

8.3.4.3 *Northeast coast*
- From 1983 to 1988, reclamation of 276 hectares of land at Punggol;
- In 1993 reclamation of 960 hectares of foreshore at the Northeast coast of Punggol;
- From 1985 to 1990, reclamation of 472 hectares of land at Northeast Coast (Phases 1–3) between Seletar and Punggol at a cost of US$180 million;
- Beginning in 1996, reclamation of 15 hectares off Coney Island; thereby doubling its size. Half of the island was a park, 1,400 low-rise homes were built on the other half;
- In 1997, reclamation of approximately 40 hectares of foreshore between Punggol and Coney Island involving approximately 11.0 million m^3 of fill;
- From November 1997 to 5th November 2001 reclamation of Phase 4 of the Northeast Coast of Punggol for private residential developments interspersed with public housing, commercial and recreational developments. The work involved reclaiming some 155 hectares of land from the foreshores of Punggol and Coney Islands, costing about US$152 million. The reclamation enlarged Coney Island and created a new water channel varying from 100 to 200 m wide between Punggol and Coney Island (Figure 8.7).

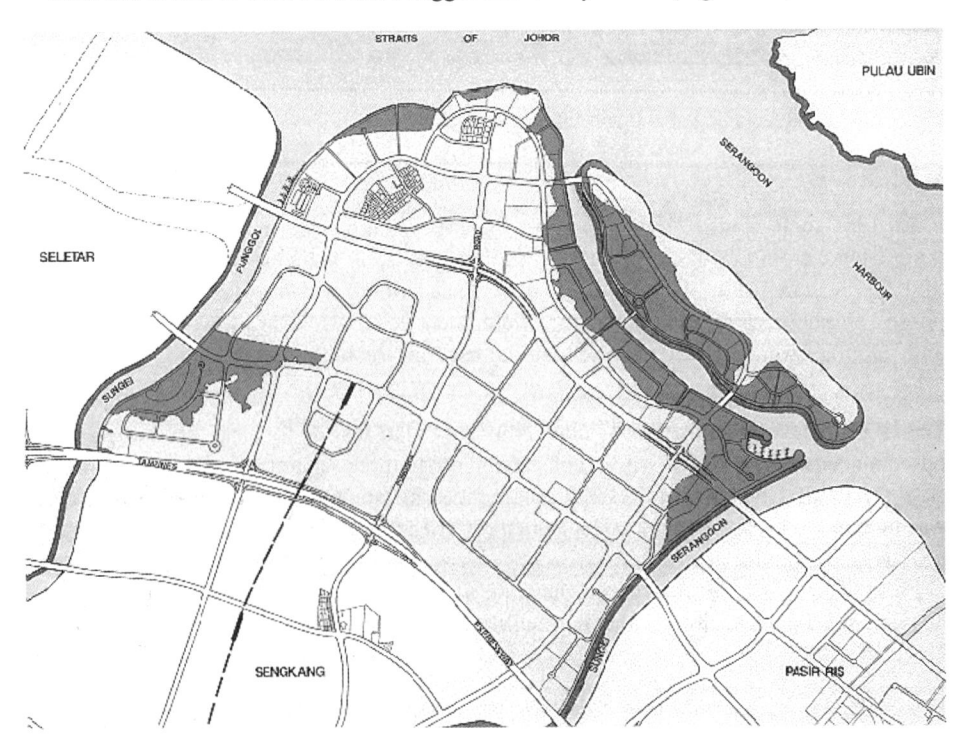

Figure 8.7. Land reclamation at Punggol and Coney Island.

- The Phase 4 reclamation included placing about 10 million m³ of sand. The average fill depth was 6.5 m, total granite stone quantity is 456,000 m³. Dredging and backfilling of a trench amounted to about 3.5 million m³. A 8.67 km retaining wall and revetment works were constructed. The main contractor was a joint venture of TOA Corporation and Jan De Nul NV.

8.3.4.4 Woodland Checkpoint
From 1990 to 1992, reclamation of 9.7 hectares in the northwest of Singapore for the construction of Woodland Checkpoint.

8.3.4.5 Kallang Basin
Reclamation of 199 hectares of land at Kallang Basin.

8.3.4.6 Tuas checkpoint
Between 1993 and 1995, reclamation of 20 hectares of land from the sea off Tuas, at the west end of Singapore, for the construction of a customs, immigration and quarantine facility for the second bridge crossing Singapore-Malaysia.

8.3.4.7 West coast
Reclamation of 86 hectares on the west coast of Singapore.

8.3.4.8 Pasir Ris
Reclamation of 44 hectares.

8.3.4.9 Marina Bay and Tanjong Rhu
From 1992 to 1995 reclamation of 38 hectares at Marina Bay and 5.6 hectares at Tanjong Rhu.

8.3.4.10 Pasir Panjang
Reclamation of 4.7 hectares at Pasir Panjang on the west coast of Singapore.

8.3.4.11 Southern island reclamation
May 2000 and March 2003, reclamation of about 35 on the foreshores of Renget Island, Lazarus Island, Renget Shoal and Kias Shoal (Figure 8.1) on behalf of the Singapore Tourism Board (STB). Three causeways have been constructed to link the islands. The project cost about US$38 millions.

8.3.5 PSA land reclamation projects

Reclamation projects conducted by the Port of Singapore Authority have mainly been for seaport development. On the mainland, the airport and seaport were constructed at Changi. The development started in the early 1970s and took about 30 years, producing a total of 2,200 hectares of land. Another mainland reclamation is the Pasir Panjang Container Terminal Extension. The terminal is to be built in four phases over the next 30 years, achieving a final capacity of 36 million teus (UNESCAP website. 2003).

Apart from mainland reclamation, PSA also carried out reclamation works at more than 20 offshore islands mainly using material dredged from seabed.

8.3.5.1 Changi Reclamation
About 700 hectares of land was reclaimed at Changi, at the east end of Singapore in the late 1970s for the construction of Changi Airport. The project involved dredging and hydraulic filling of 40 million m³ of sand by cutter suction dredgers (Chao, 1980).

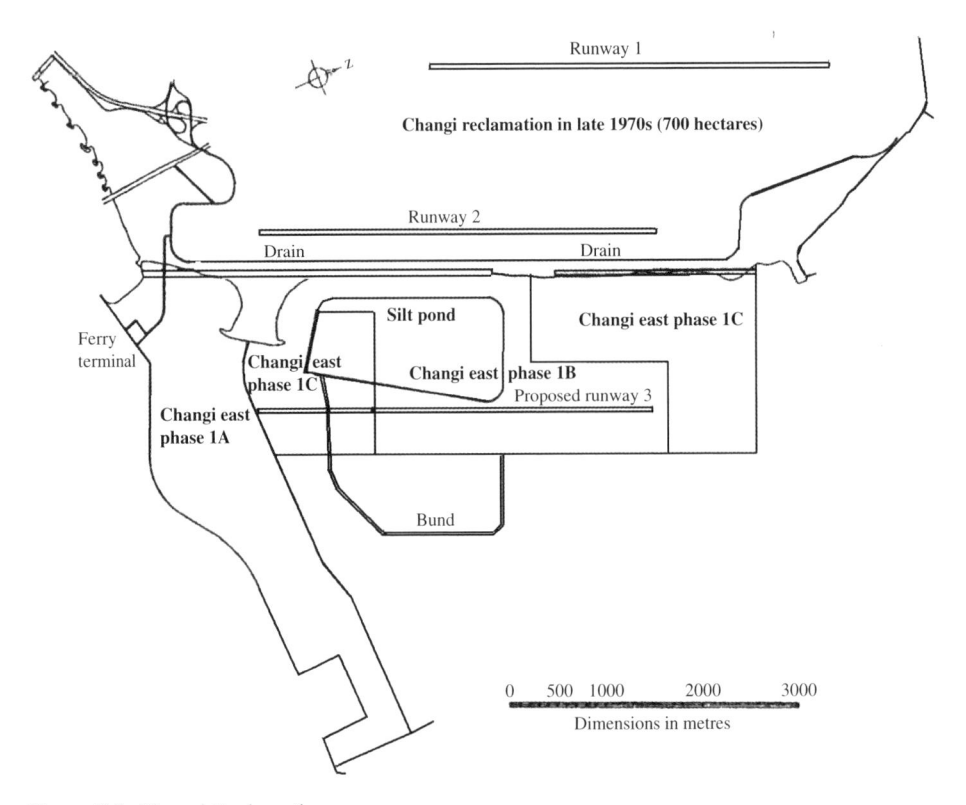

Figure 8.8. Changi Reclamation.

8.3.5.2 *Changi East Reclamation*

A further 2,000 hectares was reclaimed for the Changi East Project to cater for future expansion of Changi Airport and for a naval base, using about 230 million m^3 of sand (Figures 8.2 and 8.8). The total reclamation and dredging work cost US$1.5 billion.

Site investigation revealed that the site is underlain by the Old Alluvium comprising of a cemented silty clayey sand. This deposit is in turn overlain by the Kallang Formation comprising of soft to firm silty clays and sands of the late Pleistocene and recent deposits which are of marine and estuarine origin. The infilled valleys of marine clays are up to 40 to 50 m deep and had fairly steep side slopes. The marine clay deposits are of different ages generally separated by reddish silty clays and peaty clays caused by exposure of the seabed to the atmosphere during the rise and fall of the sea level in the geological past.

Phase 1A

Phase 1A commenced in January 1992 and finished in 1997 (Figure 8.8). A protective arm was formed so that the subsequent phases of reclamation could be carried out in relatively calm waters. About 550 hectares of land was reclaimed with about 63 million m^3 of sand. The sand was transported from mainly two borrow sources about 20 to 40 km away by trailing suction hopper dredgers as well as hopper barges for direct placement at the reclamation site, where the seabed was generally between −5 m CD and −10 m CD (Chart Datum is 1.6 m below mean sea level). This direct placement raised the level to −3 m CD. The remaining reclamation, up to +5 m CD, was carried out by hydraulic placement using cutter suction dredgers. Where the seabed was shallower, sand was placed hydraulically through 600 mm diameter pipelines.

Phase 1B
Phase 1B commenced in 1993 and was completed in 1998 (Figure 8.9). Substantial soil improvement works were involved as the reclaimed area was underlain by deep deposits of soft marine clays. One problem was a borrow pit where material had been taken from for reclamation of Changi Airport in the 1970s. This 180 ha area, known as "The Silt Pond" had been filled by silt clay washings from a sand quarrying operation leaving a depth of up to 20 m of very soft material. It was enclosed, capped and improved.

Phase 1B involved reclaiming 490 hectares of land with about 76 million m^3 of sand. The sand was transported from borrow areas about 30 to 45 km away by trailing suction hopper dredgers and self-propelled hopper barges of 9,000 m^3 and 3,000 m^3 capacity respectively. The sand was deposited into re-handling pits adjacent to the reclamation site from where it was dredged by cutter suction dredgers and pumped through 600 mm diameter pipe lines to the reclamation site. The seabed is generally between −3 m CD to −5 m CD and the reclamation generally brought the level up to +4 m CD. The Boskalis dredgers Prins der Nederland and Mark III and the Jan de Nul flagship JFJ De Nul conducted the dredging works as sub-contractors for main contractor Hyundai (DPC, 1994).

Phase 1C
Phase 1C started in 1996 and finished in 2000 and cost approximately US$55 million. It involved the reclamation of approximately 520 hectares of foreshore just beyond Changi Airport. 65 million m^3 of fill was used for the reclamation. 6.3 million m^3 of good earth and a further 3 million m^3 of soft clays in a staging ground at Changi East, all excavated from mainland projects, were brought to and subsequently transported to the reclamation areas or to the offshore disposal area (WDMC, 2001).

Trenches were dredged along the edge of the reclamation using grab dredgers with a bucket sizes ranging from 8 to 23 m^3.

Material dredged from the seabed offshore of neighbouring countries was placed at either the reclamation area or at the offshore disposal site. With the completion of the dredging of the trenches, refilling with sand started in several stages:

- Placement of a sandkey up to seabed level;
- Construction of a retaining bund around the areas to be reclaimed by means of dumping and hydraulic filling the final level; and finally
- Filling of the areas to be reclaimed.

Because soft dredged material was to be placed at the reclamation areas, the subsequent filling with sand required special techniques that allowed the sand to be placed carefully without causing disturbance of the under laying layers. For this purpose special spray-pontoons were designed and built by Jan De Nul and TOA.

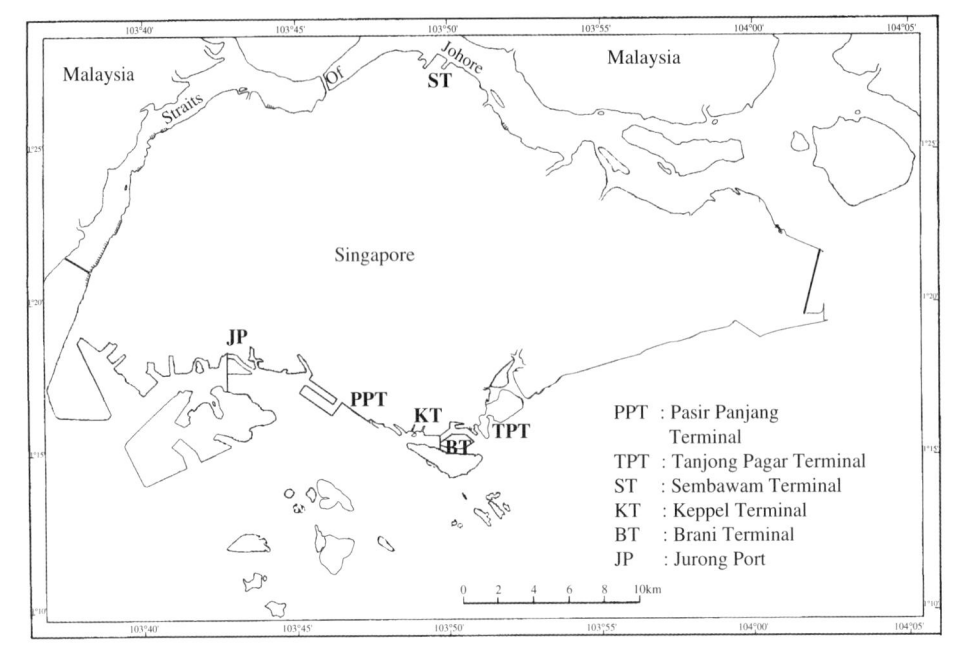

Figure 8.9. Map of Singapore showing the location of the six port terminals.

8.3.5.3 *Reclamation of offshore islands*
By 1987 PSA had carried out reclamation works at 21 offshore islands for various purposes, including catering for recreation needs (Pui, 1986).

8.3.5.4 *Reclamation and infrastructure works at the southern islands*
From June 1997 to May 1999, the PSA reclaimed about 34 hectares over the foreshore and reefs in the Southern Island to pave the way for a marine village for recreational and mooring facilities and water front housing (Raziz, 2000). Two causeways were constructed linking the islet of Rias and the 3 islands of St. John's, Lazarus and Renget (Figure 8.1).

The reclamation, costing about $144 millions involved approximately 1.5 million m^3 of sand fill, 1.0 million m^3 of rock fill and 250,000 m^3 of good quality beach sand.

Other work included the construction of approximately 5.0 km of vertical seawalls/sloping stone revetment seawalls, ranging from 6 metres to 8 metres in height plus the construction of approximately 800 metres of beach between Lazarus and Renget Islands.

8.3.5.5 *Phases of port development*
As one of the world busiest ports, Singapore has an excellent infrastructure and strategic location at the crossroads of major shipping routes, the Strait of Malacca, which connects the Pacific Ocean and the Indian Ocean. At present, the port operates six terminals, Tanjong Pager, Keppel, Pasir Pangjang, Sembawang, Jurong and Brani (Figure 8.9).

The Jurong Wharves were established in 1965. Conversion of the former British Naval Base in the north into Sembawang Terminal was undertaken in 1971. Pasir Panjang Terminal (PPT) was commissioned in 1974. The construction of the first container berth at Tanjong Pagar started in the late 1960s. Since then, the container handling facilities have expended at the two container terminals Tanjong Pagar Terminal and Keppel Terminal. A third container terminal at Pulau Brani was planned in the mid-1980s and became fully operational in 1995.

Pasir Panjang extension phase 1
Singapore started to build a new container port off the existing port of Pasir Panjang in the early 1990s, set to become the world's largest port. The terminal was to be built in four phases over the next 30 years, achieving a final capacity of 36 million teus (UNESCAP, 2004).

From 1993 to 2000, PSA spent US$210 million on the reclamation along the foreshore of the west coast of Singapore and construction works for Phase I of the container terminal at Pasir Panjang. The work included reclamation of 129 hectares of land and the construction of eight container berths with a total length of 2,730 metres. 9.3 million m^3 of seabed sand were placed in the Phase I site by Dutch contractor HAM (IHC, 2003).

Generally, the Phase 1 site consists of sedimentary rocks of the Pasir Panjang-Jurong Formation, which was formed by consolidation by overlying sediments and/or cementation of sedimentary deposits through geological time.

The overlying layer above the bedrock, which consists of sedimentary deposits, ranges from very soft marine clay at the top and alternating layers of sand and firm to hard silty clay below the marine clay. Just above the bedrock is decomposed rock. The thickness of the over laying layers is highly variable with thickness varying from a few metres to more than 30 metres.

The underlying bedrock ranges from siltstone to very strong slightly weathered limestones. The limestones appear to be interbedded with sand stones, siltstone, mudstone and shales.

Pasir Panjang extension phase 2
Phase II of the development started in 1996, reclaiming 219 hectares to the west and south of Pulau Retan Laut; and dredging and delivering16 million m^3 of sand for construction of the container terminal. Simultaneously the Pasir Panjang Terminal was deepened to 15 m, a 17 million m^3 project conducted by Penta Ocean and Boskalis, while Hyundai and Koon Construction of Singapore are building 2.5 km of quay and reclaiming land. Phase II will provide 18 berths and is due for completion in 2009. The main contractor is HAM and the contract is valued at US$38.5 millions (HAM, 2003).

The site is generally underlain by siltstone, sand stone and limestone in various degrees of weathering ranging from residual soils of stiff to hard clayey silt/silty clay. The weathered rock is more intact at greater depth. Some shallow sandy and soft marine clay deposits, including filled limestone cavities, were found during the site investigation. Unlike the Phase I site, the marine clay deposits were confined to a few small areas particularly in the north part near the island and south-eastern part of the project site.

8.3.6   Dredging for navigation channels

8.3.6.1   *Dredging of rock shoals at Main Strait*
In 1980, the Jun De Nul Group dredged rock shoals at Main Strait for the Port of Singapore Authority; costing US$5.95 million.

8.3.6.2 *Brani terminal: approach channels and basins*
In 1989–1990, the Jan De Nul Group carried out the deepening project for the Port of Singapore Authority costing US$23.6 million.

8.3.6.3 *Singapore harbour: dredging of approach channels, fairways and basins*
In 1987–1988, the Jan De Nul Group carried out the project for the PSA costing US$88.9 million.

8.3.6.4 *Maintenance dredging at Tanjong Pagar Terminal and approaches*
From 12 May 2003 to 11 August 2003, maintenance dredging was conducted at Tanjong Pagar Terminal and its Approaches.

8.3.7 Dredging for pipeline laying

8.3.7.1 *Deep tunnel sewerage system*
Dutch dredging contractor Royal Boskalis Westminster acquired a substantial order from the Ministry of the Environment for the construction of a large outfall in the Changi area (DNO, 2002) (Figure 8.10). The order, valued at approximately €135 million is being conducted in co-operation with Archirodon. It started in June 2002 and is expected to finish by May 2005.

The project entails laying three pipelines for the disposal of treated effluent from the new Changi Water Reclamation Plant. The work includes dredging a trench offshore, laying a rock mattress for the pipelines, installation of the pipelines onshore and offshore,

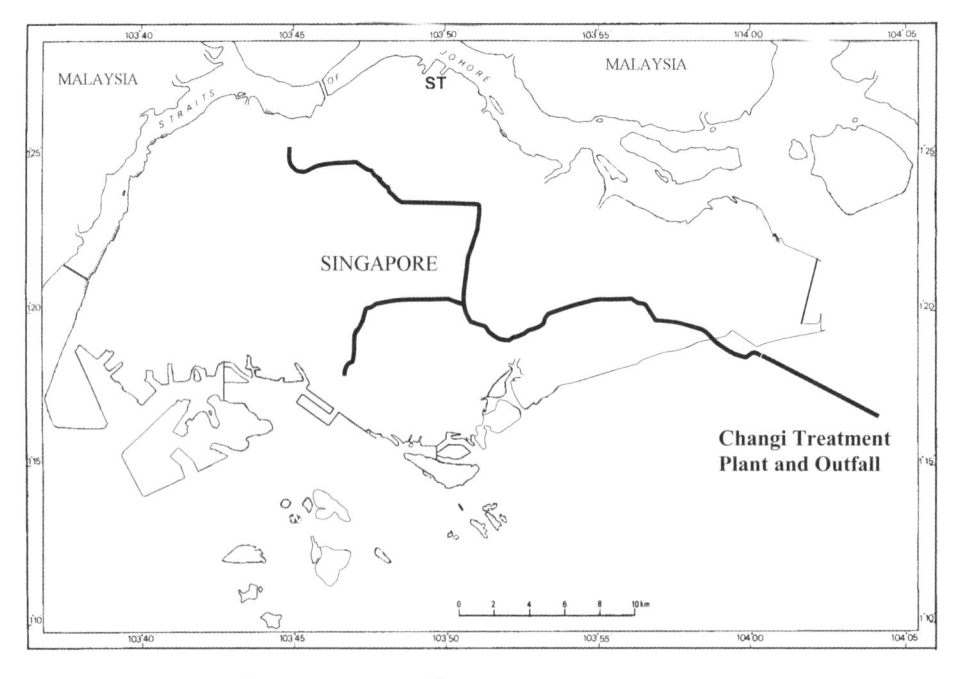

Figure 8.10. Deep tunnel sewerage system, Singapore.

and covering the pipelines with protective sand and rock. The agent for the Singapore government is the Ministry of the Environment.

8.3.7.2 *Installation of offshore pipelines from Indonesia to Singapore*
A number of projects related to installation of offshore pipelines was awarded to Jan De Nul with a total value of more than US$130 million (DNO, 2002). The works involve three sections of the gas pipeline from Indonesia to Singapore. Jan De Nul is the main contractor. Trench dredging in hard soil and more than 2 million tonnes of rock placement for pipeline protection form the major part of these works.

8.3.7.3 *West Natuna transportation system*
Pertamina (Indonesia's National Oil Company) and Sembawang Gas (Singapore) entered an agreement to supply gas from the West Natuna gas fields to the industrial complexes of

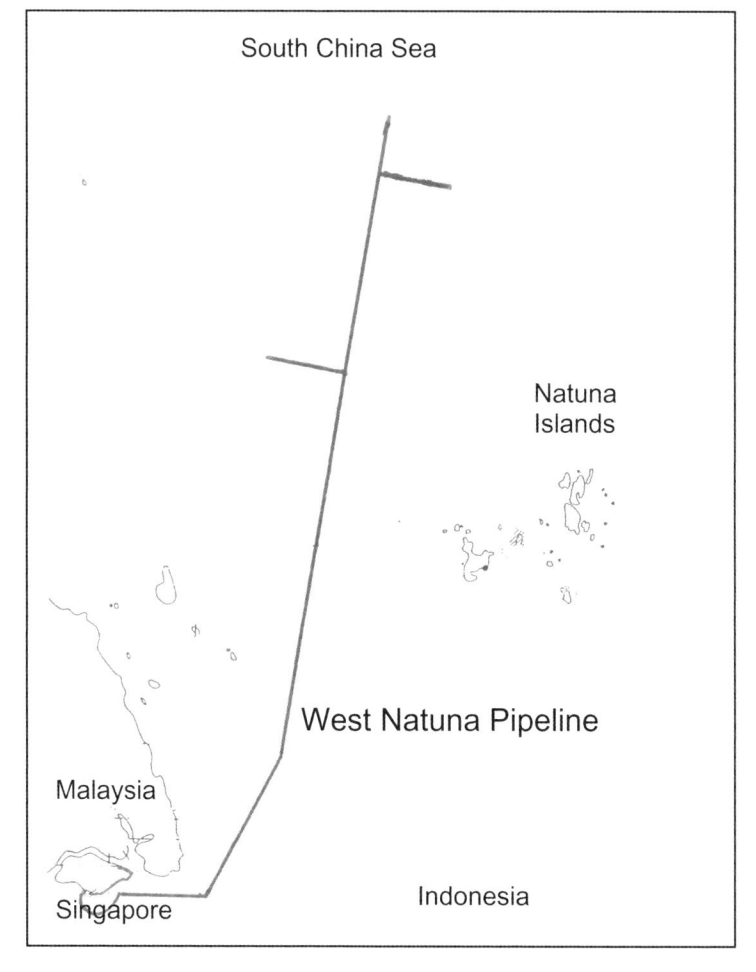

Figure 8.11. West Natuna transportation system.

Singapore. The 600 km long, 28" diameter, high pressure trunkline was laid from Singapore to West Natuna Oil Field in the Indonesian Continental Shelf (Figure 8.11).

The pipeline was laid over very rough seabed with irregular subsoil conditions (coral and rock outcrops). A significant amount of seabed rectification works was executed by way of dredging and rock placement. More than 2 million tonnes of rock had to be installed within 100 days. Furthermore a landfall had to be constructed. Boskalis equipment was used for the dredging and rock installation works for the main contractor PLD and indirectly for McDermott.

The total scope of pre-lay rock support work comprised the installation of over 130 rock supports, ranging in height from 0.3 m to 6 m and a total of 30,000 tonnes of rock.

The pipeline needed to be protected from anchor drop and drag in the Singapore Port and International Shipping lanes with an armour rock berm. A total length of 21.2 km of pipeline was covered, using 2.15 million tonnes of rock.

8.3.7.4 Shell Pandan pipeline laying
In 1990, Jan De Nul Group dredged the trench (partly in rock) for pipelines for the client Sembawang Mar. Ltd. The dredging cost was US$3.84 million.

8.3.7.5 Powergas submarine pipeline project
From 1999 to 2003, Jan De Nul Group conducted the project for Powergas Ltd. including the engineering, procurement and installation of a 9.6 km 28" submarine pipeline and associated drench dredging, blasting, rock armour protection, coating and landfall cofferdam installation.

8.3.7.6 SMB-trench dredging of pipeline
In 1980, Jan De Nul Group dredged the trench and backfilled a pipeline for the Port of Singapore Authority and Shell costing US$4.66 million.

8.4 IMPACT OF DREDGING ON THE ENVIRONMENT

8.4.1 Impact on coral reefs

Singapore is endowed with considerable biological wealth. An early hydrographic chart showed that fringing and patch reefs grew around both the main island and more than 60 small offshore islands (Burke et al, 2002) (Figure 8.12). These reefs contain more than 197 hard coral species in 55 genera and 111 species of reef fish from 30 coral families. Singapore's coral reef area was estimated to be about 54 km^2 (Dikon, 2001).

Figure 8.13 shows that fringing reefs extended along the south coast of the main land of Tuas, at the western entrance to the Straits of Johor, to Keppel Harbour. Reef flats attained widths of 400–600 m, with a maximum width of 1000 m in some embayments (NSS, 2003). The main island is highly developed, and large areas of fringing reefs have been destroyed by land reclamation.

The offshore waters are believed to have become turbid with high suspended sediment concentrations due to the reclamation and dredging activities, threatening the remaining coral reef communities (NSS, 2003). Visibility has reduced from 10 m in the 1960s

(Chua and Chou, 1992) to less than 2 m now (Chou, 2001). High sedimentation rates of 44.64 g cm^{-2} day^{-1} have been recorded (Low and Chou, 1994) and input of sediment into the marine environment continues from several large-scale, on-going reclamation and development programmes. These include the extension of Changi coast at the eastern end, and the development of cargo container facilities along the south-western coast of the mainland. In the southern islands, the expansion of Sentosa Island, the merger of the Ayer Chawan group of islands (to create a single "Jurong Island"), the southwest land reclamation at Tuas View, and the creation of a land fill on the eastern side of Pulau Semakau have added to sediment loading of the seas.

By 1987, around 60 per cent of total coral reef had been lost to nearshore reclamation, and the accompanying sediment loads have triggered declines in coral cover. Most reefs reduced by up to 65 per cent between 1986 and 1999 (Spalding et al, 2001). There is no national policy or specified agency in Singapore to manage coral reefs.

8.4.2 Impact on sandy mudflats and seagrass

The Nature Society reports that inter-tidal sandy mudflats and seagrass beds are fast disappearing due to reclamation and dredging. A reprieve for Pulau Ubin has saved an extensive patch for the time being but the other equally extensive patch at Pulau Tekong is in imminent danger of being wiped out by reclamation (NSS, 2003).

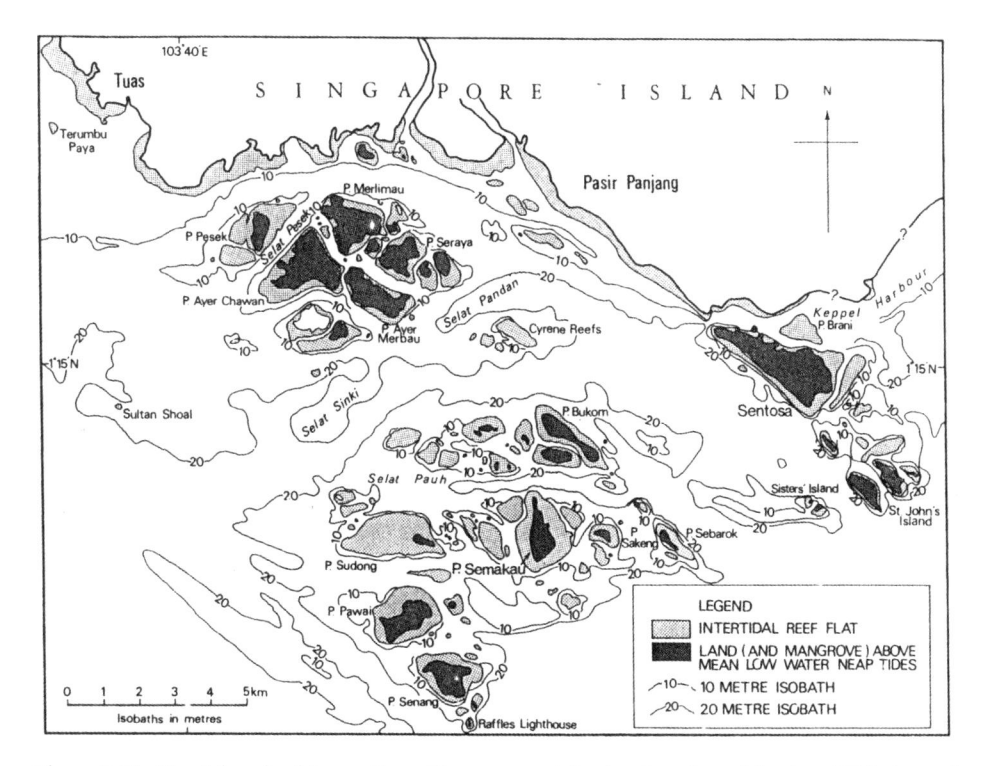

Figure 8.12. Coastal reefs of the southern Singapore mainland and southern islands, c 1953. Area of intertidal reef flat approximates the extent of reef deposits.

8.4.3 Impact in Punggol river

Punggol estuary is located on the northeast coast of Singapore. Its mangrove fringed habitat had been highly impacted by anthropogenic activities such as reclamation, dredging, clearing of mangroves and placement dredged material. These activities have resulted in the input of various pollutants into the system, resulting in an overall imbalance in the resident biotic communities (Nayar et al, 2000).

A three-year annual survey (1998–2000) at Sungei (River) Punggol, designed to investigate the effects of coastal reclamation on macrobenthic community concluded that coastal reclamation had a damaging effect on macrobenthic infauna beyond the reclaimed area. Very few benthic animals were found near the reclamation area.

8.4.4 Protection measures

8.4.4.1 *Environmental constraints*

In recent years and with increasing awareness of the need for a balance between industrial development and environmental prevention, the dredging and reclamation industry has been subjected to critical review of the environmental impact of its operations. The awareness of environmental impact has moved from the immediate and obvious, such as noise, vibration and smell, to the less obvious and often difficult to measure, such as changes in water and soil quality, movement of contaminating chemicals, slow alteration to site ecology, siltation of the approach channels and erosion of the existing land and seabed.

There are three main areas of potential impact in any reclamation project, the dredging site, the transportation route and the reclamation site. It is important to identify and mitigate any undesirable environmental impact. Impacts due to the action of dredging and disposal can be mitigated by selection of an appropriate type of dredger and method of disposal. Substantial investigations have been made at the design and planing stage to assess potential impacts related to the location and size of the dredging and reclamation sites.

Computer models are used to simulate the effect of reclamation on the surrounding water in terms of tidal flow patterns, water level, sedimentation and water quality. Approval from Parliament is dependent on the outcome of such studies.

8.4.4.2 *Protection measures*

Measures have indeed been taken to lessen the negative effects on the environment. For example, for the construction of Pulau Semakau, before dredging and reclamation work started, a survey was carried out to establish whether there were live corals within the proposed site. During construction, silt screens were erected to keep the sediment within the working area. The water quality was monitored regularly. The Nature Society recommends that when dredging sea channels, silt should not be thrown to both sides of the lane but should be carried away by sand barges (NSS, 2003).

8.4.4.3 *Measures against illegal disposal*

The reef community as a whole faces the threat of indiscriminate disposal of soil from excavation and dredging, as well as illegal poaching. For example, although the Port of Singapore Authority has set aside a disposal site east of Pulau Semakau, it is reported that some barge drivers did not follow the rules and discharged the material while the barge was moving, spreading it around. To combat this, every craft transporting material for disposal is now required to be fitted with a device which records the vessel's draught while it is en route to the dumping ground. Between 1991 and 1996, more than six million cubic metres of spoil have been placed at the official site.

8.5 DISPUTES WITH NEIGHBOURING COUNTRIES

8.5.1 Dispute with malaysia

The following information has been obtained from various press releases and website news items. The author has not attempted to verify the justification of claims and counter claims but simply reported them because it is a major concern at the present time and the outcome may have a significant effect on the future reclamation plans stated in the first section of this chapter.

There are many specific claims and it is not appropriate to list them all but rather to note the basis of the complaints which includes:

Fishing
- Loss of fish breeding grounds due to removal of the sea bed sand;
- Loss of nutrients, particularly for the dugong fish;
- Interference with fishing activities by sand barges;
- Damage to fishing nets by loose coral and debris arising from dredging activities;
- Reduced fishing areas caused by recreation;
- Narrowing of navigation channels caused by reclamation, causing dangerous mixing of fishing boats and large sea-going vessels;
- High turbidity caused by dredging and reclamation activities.

Safety of navigation
- Narrowing of channels by reclamation;
- Siltation of channels caused by sediment release from reclamation and dredging operations;
- Sand barges operating in main sea channels;
- Changes in current patterns due to deepening and to reclamation works.

Coastal processes
Changing the shoreline and the depth of water adjacent to it changes the wave climate and the currents. This in turn can cause:

- Areas of accretion
- Areas of coast erosion

Legal issues
Some of the above issues are addressed by the legal permitting framework so some of the complaints arise simply from the illegality of operations such as:

- Taking sand from unlicensed areas
- Disposing of unsuitable material in unlicensed areas

Feelings run high on these issues, particularly when people's livelihoods are threatened. There have been cases where dredging plant has been impounded by village people and people have even been killed and injured in incidents.

Some of the impoundings have been official and dredging companies have been asked to pay very large sums of money for their release. Dredging companies have invested heavily in large purpose built vessels and can ill afford to have them idle.

Financial issues

Given the high demand for sand in Singapore, Malaysia and Indonesia have been receiving and are set to receive income from their offshore sand resources. Some of the problems associated with that are:

- Agencies have negotiated rights then taken advantage of the demand by raising prices as much as tenfold;
- There are discrepancies between records of imports and exports of sand;
- International NGO's complain that the money earning potential is of greater interest than environmental conservation.

Efforts to resolve these issues have not met with a lot of success so far and it remains to be seen how much future reclamation plans will be affected.

8.6 FUTURE DEVELOPMENT

Singapore has realised that if she continues to reclaim, she will eventually infringe her neighbours' waters. In the meantime she encounters many practical problems before the territorial limit is reached. According to the 1982 United Nations Convention on Law of the Sea, every state may claim territorial waters not exceeding twelve nautical miles from the coastal baseline. Regionally, Singapore has rights within three nautical miles of its shoreline, however, due to the close proximity of some parts of the coastline to her neighbours, the territorial rights of some areas such as the Straits of Singapore are shared.

Another consideration is the depth of water. Areas deeper than 15 m are uneconomic to reclaim and there is a competing need for sea-lanes and anchorages. However, tests are underway to see if it will be possible to carry out a further 30 m reclamation at Changi, where the water is up to 40 m deep.

Current and future large land reclamation projects

Land reclamation has substantially modified the coastline of Singapore, extending it seawards, especially on the eastern, north eastern and western parts of the island, changing it beyond recognition. The coastline has also been straightened by building dykes across estuaries, particularly in the west coast across the estuaries of Tengeh, Poyan, Murai and Sarimbun. Many offshore islands have been enlarged. The Singapore Future Plan 2001 shows yet more changes, demonstrating how modern dredging has the capability to quickly and significantly change the map of the world.

REFERENCES

Bogaars G (1956). The Tanjong Pagar Dock Company (1864–1905), Memoirs of the Raffles Museum, No. 3.
Burke L, Selig E and Spalding M (2002). Reefs at risk in Southeast Asia. World Resources Institute.
Chao V (1980). Geological aspects of a hydraulic fill reclamation project. Proc. 6th SE Asian Conf. Soil Engng, Taipei 1, 469–484.
Chen CN and Sun W (1999). Numerical study on mitigation measures to the formation of sand bars at the mouth of Changi Airport Diversion Drain at east coast of Singapore: Technical Report, Nanyang Technological University.
Chou LM (2001). Country Report: Singapore. International Coral Reef Initiative (ICRI).
Chuah SG and David TTL (1995). Reclamation of Jurong Island, Engineering for Coastal Development, Nov 27–28 1995, Port of Singapore Authority, pp. 111–119.
Dikon A (2001). Effects of sediment load on coral reefs of Singapore, PhD thesis, Dept. Biological Science, NUS.
DNO (2001). SLM to dredge Indonesian Sand; Dredging News Online, Vol. 1, issue #53, 11 May, 2001. http://www.sandandgravel.com.
DNO (2002). Boskalis acquires substantial order in Singapore; Dredging News Online, Vol. 1, issue #80, 17 May 2002. http://www.sandandgravel.com.
DNO (2002). Offshore projects awarded to Jan De Nul; Dredging News Online, Vol. 1, issue #89, 19 September 2002. http://www.sandandgravel.com.
DPC (1994). Article in Dredging + Port Construction, October 1994, pp. 47.
DTE (2001). Sand mining destroys community resources; Down to Earth No. 51, November 2001; http://dte.gn.apc.org/51snd.htm).
Government of Singapore (2001). Concept Plan 2001; http://www.ura.gov.sg/conceptplan2001/
HAM (2003). http://www.hamdredging.com/projects/hamp8.htm.
HDB (2004). Review of HDB's Activities for FY 1996/97; Statistics & Charts-Cumulative Achievements (Since 1960) http://www.hdb.gov.sg
IHC (2003). Enlarging Singapore, dredgerwise; http://www.ihcholland.com/B_ihc_ dredging/ B08/b08.10_land_reclamation/b08.10_enlarging_singapore.htm.
Lee SL (1987). Layered Clay-Sand Scheme of Land Reclamation, Journal of Geotechnical Engineering, Vol. 113, No. 9, September 1987, pp. 984–995.
Leung CF, Wong JC, Manivanann R and Tan SA (2001). Experimental Evaluation of Consolidation Behavior of Stiff Clay Lumps in Reclamation Fill. Geotechnical Testing Journal, GTJODJ, Vol. 24, No. 2, June 2001, pp. 145–156.
Moch. N. Kurniawan and Tiarma Siboro (2002). House Calls for Sand-mining Ban as Marine Damage Looms. The Jakarta Post, Jakarta, Indonesia; September 10, 2002.
Nayar S, Goh B and Chou LM (2000). Nutrient and Biotic Fluxes in Relation to Dispersal of Pollutants in Ponggol Estuary, Singapore. 2nd UNU-ORI Joint International Workshop for the Marine Environment; Coastal Ecology, Nutrient Cycles and Pollution; 3–8 December 2000 – Otsuchi, Japan.
Ng P, Law WKH, Toh AC, Tan GP and Lau AH (1995). Development of the New Container Terminal at Pasir Panjang; Engineering for Coastal Development, Port of Singapore Authority, Nov 27–28 1995, pp. 81–96.
NSS (2003). Feedback on The Singapore Green Plan 2012. Nature Society of Singapore; http//www.nss.org.sg/conservationsingapore.
Pui SK (1986). 100 Years Foreshore Reclamation in Singapore; 20th Coastal Engineering Conference, 1986, Taipei, Taiwan, pp. 2631–2636.
Radhakrishnan R, Ng FW, Rajendra AS and Pui SK (1983). Land reclamation for Singapore Changi Airport. Proc. International Land Reclamation Conference, 1983, Essex, England.
Raziz Capt. HRA (2000). Dredging and Reclamation Opportunities in South East Asia. 2nd Asian and Australian Port and Harbours Conference, Malaysia, 16–18 April 2000.
Shankar NJ and Sun W (awaiting publication). Stability analysis for beaches at the southeast coast of Singapore; Journal of Coastal Engineering, waiting for publish.

Spalding MD, Ravilious C and Green EP (2001). World Atlas of Coral Reefs. Prepared at the UNEP World Conservation Monitoring Centre. University of California Press, Berkeley, USA.

UNESCAP (2004). *www.unescap.org/tctd/pubs/files/reviews2001_portsinvest.pdf*.

WDMC (2000). Dredging Transforms Singapore; World Dredging, Mining and Construction, November, 2000.

WDMC (2001). Singapore Operation, Jan De Nul NV. Article in World Dredging, Mining and Construction, June 2001.

Wei J and Chiew LK (1983). Performance of Artificial Beaches at Changi; Proc. International Land Reclamation Conference, Essex, England, pp. 67–100.

Chapter 9

Dredging in a Changing Environment

Pier Vellinga
Dean of the Faculty Earth and Life Sciences, Vice Rector Vrije Universiteit and Professor in Environmental Sciences and Global Change, Amsterdam, The Netherlands

9.1 ABSTRACT

Human growth in the last two centuries has been exponential. The question is how long can this continue without interfering with other species and nature itself. This is a source of many studies and is one of the greatest concerns of society. As we continue to apply our human expertise to "improving" nature, national and international political decision-making on sustainability issues becomes ever-increasingly important. Still, when we observe the day-to-day discussions and decisions on the use and protection of nature, political and industrial roles are often experienced as erratic and irrational to the actors involved. This paper will give some insights into the underlying reasons.

Globalisation provides new and better opportunities for entrepreneurs worldwide, but simultaneously it raises concerns about growing disparities in income at the national and international levels. And globalization comes with a further exploitation and degradation of the quality of life support systems, such as biological diversity and the global climate. These are issues that concern all of us.

This paper explains the concept of Coevolution which promotes benefits, both for the human species and ecosystems and natural habitats in design, operation and staffing; and the "Triple P Performance" principle, i.e. Profit, Planet and People as a guideline for modern companies to create legitimacy for growth amongst clients, shareholders, employees and society at large.

The paper was the keynote address at the CEDA Dredging Day, "Dredging Seen, Perspectives from the Outside Looking In", in Amsterdam, November 15, 2001 and first appeared in its proceedings. It is reprinted here in an adapted form with permission from the author and the Central Dredging Association.

9.2 INTRODUCTION: ENVIRONMENT AS AN ISSUE

The human species is colonising the earth in a successful way. Our ancestors felt vulnerable with regard to the forces of nature and they felt dependent on the gifts of nature. It is not surprising that nature is a major focus in all early religions: Amon Ra, Gaia, Wodan and so on.

Over the last 10,000 years the human species has grown rapidly in comparison with the 3 million years before. Over the last 200 years the growth in numbers has been exponential.

Figure 9.1. Man-made wetlands: a way in which the human species tries to restore a natural environment and protect nature against the loss of other species.

The main reasons for growth are:

- Progress in the human capabilities to exploit nature through the domestication of plants and animals;
- Progress in understanding the relationships between water quality and public health;
- Progress in protecting ourselves from the forces of nature; and
- In the efficient use of nature's resources, e.g. fossil fuels, water and increasingly the gene pool.

But, every now and then people ask themselves: how long can a species grow exponentially in numbers, in particular when such growth comes with the increasing loss of other species? This question together with the history of being very vulnerable to the forces of nature makes our species presently rather insecure in dealing with nature. It should not be surprising therefore that political decision-making regarding the management, use and protection of nature is often seen as erratic and irrational to the actors involved.

The decision-making process about gas exploration in the Waddenzee in the north of the Netherlands is a typical example. Another example is the use of tunnels to avoid interference with the landscape. This article will try to shed some light on the various ways our species deals and can deal with nature (Figure 9.1).

9.3 DIFFERENT CULTURAL PERSPECTIVES ON THE VALUES OF NATURE

The perspective of people regarding nature is dependent on the conditions they are in, and such perspectives change over time as one can imagine. Still, there are some remarkably consistent differences amongst people and cultures even within Europe.

Figure 9.2. In the Gallic tradition, nature is particularly valuable once humans have transformed it into food as can be seen in this French landscape of neatly planted vineyards.

Tacitus already described the habits of German tribes regarding rituals and nature. In the Germanic culture nature is considered holy (see: Simon Schama (1995): Landscape and Memory, Alfred A. Kopf, New York, 625 pp.). This religious attitude with regard to nature can presently be recognised in the "Fundi" movement of the German NGOs when it comes to environmental issues. In the debate about genetically modified organisms (GMOs), the naturalness of reproduction and the holiness thereof is a central theme.

In the Roman and Gallic tradition nature is culture: nature is particularly valuable once humans have transformed it into a park landscape or into food. This attitude can be recognised in the Italian and French landscapes and also in the present political debates regarding farming systems and related land use management as part of the food culture in France. In the debate about the introduction of GMOs in France, the protection of the farming system and the related landscape and food culture is the central theme (Figure 9.2).

In the Anglo-Saxon tradition nature is there for use. The United Kingdom always had a tradition of free access to nature through its pathways. The Anglo-Saxon tradition is one of pragmatic use of natural resources. Regarding the GMO debate it is not completely clear, unless the USA approach is illustrative where there is little or no opposition to genetic modification.

All those involved in industrial activities and in reshaping the earth's surface in one way or another may ask themselves the question, to what extent do you recognise the cultural differences in Europe and/or elsewhere in the world when it comes to important human interference in the natural systems.

There are parallels in geographic responses today. German environmental research institutes for example were the first to include the number of tonnes of excavation and removal of soil material in the set of indicators for material intensity of the economy. This so-called MIPS indicator (Material Input per Unit of Service) is a measure to describe the environmental burden of a certain activity. It has been developed by the Wuppertal Institute and includes soil excavation and removal, even if the environmental effect is negligible.

Experts in the USA, UK and the Netherlands are hesitant to adopt the German approach. However, the discussion in the Netherlands regarding a green tax on the excavation and removal of soil including dredged spoil irrespective of its environmental quality and in respect to the environmental burden, illustrates that Fundi approaches can also be found in the Netherlands. Green taxes in themselves are a proper instrument for environmental management, but there should be a clear relation to environmental effects. The Roman, Gallic and Anglo-Saxon cultures will probably not immediately embrace such a green tax on soil removal unless there is clear environmental stress involved.

9.4 FROM PERSPECTIVES TO ATTITUDE: VIEWS ON DEALING WITH NATURE

A view on nature can be defined as a philosophical opinion on how to deal with nature. This opinion includes both an objective concerning nature and a strategy to reach this objective. Societal views on nature as they are empirically encountered can be classified into three main streams on the basis of their objectives, whilst their implied strategies allow for some refinements.

Three main views that have been identified concern (Ruijgrok and Vellinga, 1999) (Figure 9.3):

- The Conservation view;
- The Development view;
- The Functional view.

9.4.1 Conservation view

The Conservation view is mainly concerned with the objective of conserving and restoring existing natural sites according to an historical reference situation. Whether these have been

Criteria	Conservation		Development		Functional	
	Hands off	Classic	Development		Coevolution	Nature building
1. Trade off ecology and economy	Priority on ecology	Priority on conserving natural and cultural sites	Priority on naturalness		Careful trade off to maximize social welfare and natural qualities	Priority on economy
2. Active human intervention	No influence allowed	Activities to conserve species	Actions to stimulate/enable natural processes		Only if it creates win-win situations	Steering natural processes by engineering techniques

Figure 9.3. Criteria for basic views on nature.

obtained by human interventions or not is not an issue here. What matters is the protection of the nature remaining today. The protection of today's nature can be realised in two ways:

1. Through keeping one's hands off existing nature. This strategy is opposed to human influence on the natural surroundings. The belief is that human intervention always reduces the naturalness of an ecosystem. Naturalness is defined as the extent to which nature is free from human interventions. The best way to conserve nature is not to touch it and to rely on the natural restoration capacity of ecosystems.
2. Through maintenance and isolation of existing natural and cultural landscapes, aimed at protecting rare species and unique cultural and historical elements. Active human intervention is considered necessary since nature cannot defend itself against the threats from society. Intervention activities such as fencing and mowing are done to maintain diversity and rarity of species.

The Conservation view is a reactive/defensive attitude as a response to rapid industrialisation and related land use changes. As such, it may become counterproductive. Productive measures aimed at preserving diversity and rarity of species can especially reduce natural dynamics (for example succession possibilities). This gives rise to the question whether or not nature is served through such measures.

9.4.2 Development view

In the Development view both protection of existing nature and the development of new natural sites are the main objectives. Driven by the desire to enhance naturalness and wilderness, room for natural processes and diversity of systems instead of species are core issues of this view. Natural processes and dynamics are restored by stimulative actions with relatively uncontrolled spontaneous end results.

There are a number of ways to reach these objectives:

1. Reducing maintenance whenever possible to give natural processes a chance, such as managed retreat;
2. Undoing previous interventions, such as the opening of enclosed sea arms; and/or
3. The creation of abiotic conditions, which will reactivate natural processes, such as dune formation with the initial aid of sand screens.

Whether it is a reduction of maintenance, the undoing of interventions or the creation of physical conditions, they are all aimed at the enhancement of naturalness. This can be realised when connections are made between existing nature; ecological networks or corridors are stimulated since they help to enhance natural resilience of ecosystems. Basically, stimulative interventions are driven by the desire to provide more room for nature rather than by the desire to realise utility for society.

An important argument supporting the Development view is that merely conserving existing natural sites is not adequate. Increasing the quantity and quality of nature requires ecological networks and room for natural processes in addition to protection and isolation.

The creation of corridors and space for processes requires stimulative interventions in a society in which every piece of land is increasingly being used for human activities; neither protective isolation nor simply keeping one's hands off nature is sufficient to secure nature and natural dynamics.

9.4.3 Functional view

According to the Functional view, nature derives its value directly from the welfare functions it performs for society. This does not mean that nature cannot have an intrinsic value according to this view, but it will only have one if humans find it important for some reason. The objective of this view is maximisation of social welfare derived from nature. This welfare can be derived through both direct (resource extraction) and indirect (regulation processes) use, but also through the social preferences attached to its mere existence.

Because nature is supposed to serve social preferences, which vary amongst interest groups, one can define a whole spectrum of strategies to realise the objective of welfare generations. Two examples of this spectrum are:

1. Realising social welfare through constructing nature according to human wishes. Since naturalness is supposed to be an illusion, one can control and construct nature to meet social demands with the help of civil engineering. Humans can destroy nature through technology, but people can also create favourable conditions for nature by means of technology. Nature can be man made and abiotic conditions do not pose restrictions since these can be adjusted too. This approach is referred to as the "Nature Building" approach.
2. Realising social preferences through a sustainable use of nature; only user functions that do not seriously damage the natural system are allowed, such as nature friendly forms of recreation and sustainable forms of harvest. Though naturalness is considered desirable here, it does not exclude human activities, since humans are also part of nature. A balanced interaction between nature and society is advocated. Both society and nature are allowed to change and to inflict change upon each other as long as neither of them suffers serious damage, threatening its existence; it is a matter of mutual benefit. The term "Coevolution" is used here to describe this interpretation of the Functional view.

The Functional view is based on the principle that separation between ecology and economy will neither be favourable to nature nor to society in the long run, since the two are interdependent. Sustainable use of nature will be beneficial to both nature and society. Opposition to this interdependency is simply not realistic.

A critical argument against the Functional view is that it requires operational instruments and legislation to ensure a balanced trade off so that nature is protected against the "tragedy of the commons". Examples of the functional Coevolution view are natural reserves in which certain types of recreation are allowed. Recent examples of how the Functional view can be operationalised is the Dutch plan "Growing with the Sea" and the World Wildlife Fund vision regarding the extension of the Port of Rotterdam, in such a way that both the economy and the ecology benefit (Figure 9.4).

The three main views of Conservation, Development and Functionality are more than just theoretical concepts. In practice they can be used to explain:

- Motivations of economic interest groups concerning their way of dealing with nature;
- Developments in national policies and decision-making practice in the fields of nature protection and coastal zone management;
- The way in which nature is valued and accounted for in public decision-making.

Figure 9.4. The port of Rotterdam: new plans for the extension of the port are an example of the "Functional Coevolution" view in which both the economy and ecology benefit.

Period	Policy aim	View on nature
1950–1970	Protecting existing nature	Conservation
1970–1995	Developing a national ecological network	Development
1995–2000	Call for interaction of economy and ecology	Tentative → Failure
		Coevolution → Success
2000–future	Application of balanced trade alis between ecology and economy	Coevolution

Figure 9.5. Policies for nature.

9.5 VIEWS AND PUBLIC POLICIES

In order to find out whether the views can also explain trends in public policies, this section presents a brief review of the Dutch policies for nature protection and coastal zone management. In the Dutch policies for the natural environment (Figure 9.5), one can see a clear development in the lines of thinking starting from a Conservation approach, continuing with a Development approach, and recently the first attempts have been made towards an enlightened version of the Functional approach: Coevolution.

When looking at the Dutch policies for coastal protection (Figure 9.6), a similar outcome but an opposite starting point can be discerned. Starting from a single sector civil

Period	Policy aim	View
1950–1970	Single sector flood protection	No view on nature
1970–1990	Double criterion of protection and environmental neutrality	Nature building
2000–future	Multifunctional use of coastal space and resources; "Growing with the Sea"	Coevolution

Figure 9.6. Policies for coastal zones.

engineering approach of flood protection until the beginning of the 1970s, increasing awareness of environmental impacts of coastal protection works led to the adoption of the double criteria of flood protection and environmental neutrality. At this stage the aim was to minimise environmental impacts of coastal defence measures. This approach was applied up until the early 1990s. In these first two stages the coastal zone was not considered as nature, but merely as an accumulation of (physical) defence capital.

Only recently the view has been adopted that coastal defence and nature development can go hand in hand. Both policies in the field of nature and coastal protection have shifted towards the Coevolution approach.

Whether Coevolution will become the dominating philosophy for both coast and nature largely depends on the availability of trade-off instruments and legal arrangements necessary to implement the concept of Coevolution. The legislation of the concept of compensation in the EU-habitat directive may be a starting point.

9.6 GLOBALISATION, ECONOMIC GROWTH AND THE ENVIRONMENT

Many governments and international corporations promote the further opening of global markets as a way to enhance development and income levels worldwide. However, the present rate of globalisation raises concerns about growing disparities in income at the national and international levels and a further degradation of the quality of life support systems, such as biological diversity and the global climate. The challenge that global environmental issues pose to the relationship between environment and development can be illustrated on local, regional and global scales, each with its specific environmental problems.

9.6.1 Local Average Income Levels and Environmental Pressure

Figure 9.7 (Local Average Income Levels and Environmental Pressure) reflects empirical evidence that people tend to solve their local environmental problems as income increases. Growing income levels can be correlated with an improvement in the quality of the local environment.

Many cities and countries in the industrialised part of the world have experienced the situation represented by this curve, whilst cities in developing countries can still be located on the ascending segment of the curve. The rationale behind the curve is that, as income levels rise and local environmental and health problems become manifest, there are driving forces

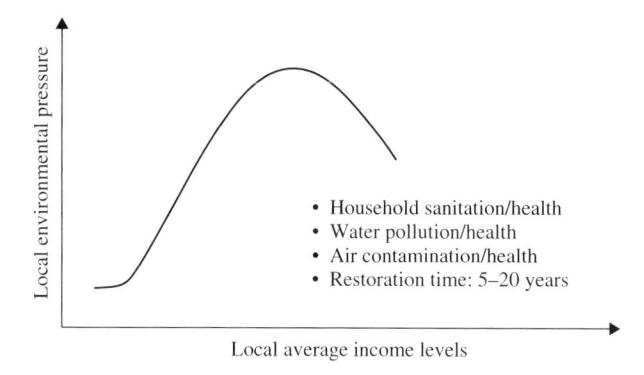

Figure 9.7. Local Average Income Levels and Environmental Pressure.

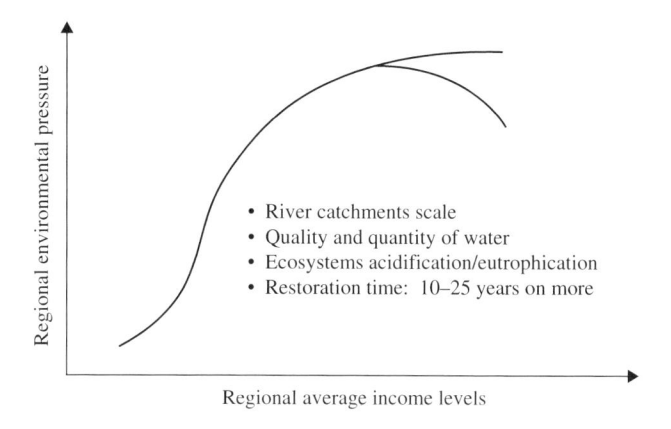

Figure 9.8. Regional Average Income Levels and Environmental Pressure.

and financial means to introduce technologies and regulations (incentives and institutions) that reduce pollution and protect the health of the population.

Two critical factors leading to success can be identified:

- People take measures when they actually see that their health is affected.
- Costs and benefits play out at the same (local/national) level.

9.6.2 Regional Average Income Levels and Environmental Pressure

A similar curve (see Figure 9.8, Regional Average Income Levels and Environmental Pressure) can be developed for environmental problems at a regional level, such as acidification and water quantity/quality issues on the scale of river catchments. There is less evidence that people address these problems successfully as income levels go up. An important reason for this is that industrial and agricultural activities higher up in river catchments (upstream and/or upwind) benefit from pollution and overuse of water and pollution of air, whilst downstream and downwind people and nations experience the negative impacts. Another reason for continued environmental degradation, as income levels go up, is the time delay between the act of polluting and the effect of pollution downstream. Examples of regions

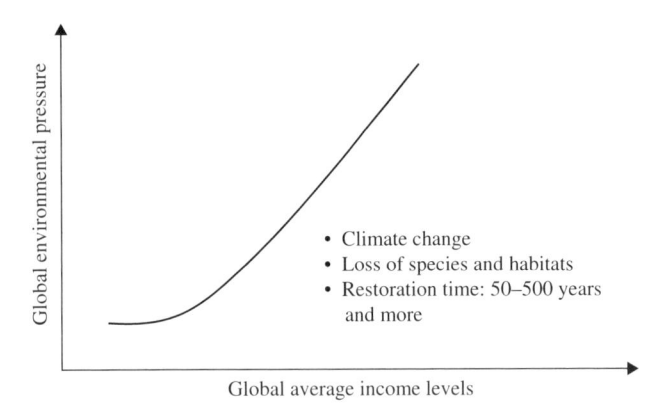

Figure 9.9. Global Average Income Levels and Environmental Pressure.

and environmental problems exist, where the curve drops sharply, but this is not a general empirical finding. For most regions of the world the evolution of the curve is not yet clear.

9.6.3   Global Average Income Levels and Environmental Pressure

A curve (see Figure 9.9, Global Average Income Levels and Environmental Pressure) can also be developed for global environmental problems such as climate change and loss of species and habitats. Empirical data illustrate that there is no income-related levelling-off point when we look at the relationship between income and emissions of greenhouse gases caused by consumption. Income levels (including the consumption of imported goods) correlate with energy use, and present-day energy use is coupled to CO_2 emissions. Similarly, the space we use for our activities (housing, transport, recreation) grows in a linear fashion to income, at the expense of natural habitats.

A critical feature of global environmental change is the time scale of biophysical response. The climate responds to changes in the concentration of greenhouse gases at a time scale of decades to centuries or even longer. The loss of species including their habitats is considered irreversible in a human time frame.

From the above observations it is clear that global environmental change poses an unprecedented challenge to society and requires a proactive approach. To avoid irreversible, high impact changes, society must act before the effects of environmental change become visible. How can research help in clarifying the issues at stake? Certainly a better understanding of the interaction between social and natural systems is required.

9.7   FROM "END-OF-PIPE" TO SYSTEMS INNOVATION

Guidance for the future may be found in analysing trends in societal response to environmental problems over the last 40 years. Figure 9.10 (from McKinsey and Company, Winsemius and Guntram, 1992) presents a number of stages of societal response to environmental problems (as quoted from Winsemius and Guntram):

1. *Reactive response*
   The onset of government policy-making is generally met with a primarily defensive approach. Companies, and especially their sector organisations, dig in exaggerated.

Development stages in corporate and societal response (adapted
from Winsemius and Guntram, 1992; and Vellinga and Herb, 1999)

 A)	Reactive	Receptive	Constructive	Pro-active
B)	End-of-pipe	Process	Product	System
C)	Specialists	Managers	Sector	Society
D)	Minimisation	Optimisation	Acceleration	Vision

A) Response phase
B) Focus of attention
C) Main actors
D) Driving philosophy

Figure 9.10. Development stages in corporate and societal response (adapted from Winsemius and Guntram, 1992; and Vellinga and Herb, 1999).

They tend to adopt a posture of loyal citizens: "We will do what is legally required, but we don't like it". Generally assigning the responsibility to staff specialists, usually as an extension to the Health & Safety departments, they implement the prescribed end-of-pipe solutions consisting of add-on features to the existing facilities whilst all the time trying to minimise their response and the costs thereof.

2. *Receptive response*
Gaining experience and becoming more comfortable with the new responsibilities, companies shift to an attitude of "Okay, if we have to do it, let's be smart about it". Line development is now made responsible – within the boundaries of the current business – for developing solutions that meet the criteria set by the government in the most efficient manner. Most solutions will still involve optimising the existing production configurations although they will often include some process redesign.

3. *Constructive response*
A limited number of companies have begun to look beyond the boundaries of their current business to find more fundamental answers to the environmental questions. In more advanced countries, stimulated by government intervention, industries have developed so-called "cradle-to-grave" approaches where they accept responsibility for their product even after it has been sold. As a result, the traditional delivery chains in, for instance, the packaging or automotive industries are changing rapidly.

Companies are also starting new cooperations with suppliers, customers, and, especially, competitors to facilitate joint objectives, such as waste collection and recycling, "green" product labelling or contractual agreements ("convenants") with governments that establish new environmental objectives. Moreover, existing approaches can often no longer meet much tighter targets, necessitating industrial players to strive for technological and/or organisational quantum leaps.

4. *Proactive response*
Very few companies have reached this phase yet. Still, the contours of the response pattern can be sketched by looking at the policy development in leading countries and at the role specific companies play. Increasingly, companies will internalise the environmental challenge as an element of quality management. To meet the challenge and, at the same

time, focus on the needs of their customers. They will optimise their own functioning and especially their value proposition (i.e., products plus services at a given cost). Companies and industry sectors will pool their resources with those of governments, scientific institutions and often environmental issues that can pass the 3E-test.

"The challenge of bridging interests and cultures among players of great diversity is considerable. Leaders in government and business, as well as in the environmental organisations and scientific institutions must generate a vision that can serve as a reference for all workers in their organisations.

"Within industry, this vision must inspire the full internalisation of the environmental challenge throughout a company. Top management must take the lead in defining far-reaching but understandable goals – DuPont and Asea Brown Boveri, for instance, talk about 'zero emissions' – that stretch the organisation to look beyond the horizon of today's concerns – for instance, 'We are responsible for tomorrow's laws'. However, building on the experience of the first three stages, this long-term vision also must drive the medium-term strategy that in turn can be build on in practical short-term action plans".

9.8 A TYPICAL DREDGING EXAMPLE OF THE FOUR STAGES OF RESPONSE

How do port authorities address the environmental aspects of dredging?

Reaction phase:
"It is not really contaminated or, the pollutants are not really hazardous to our health and ecosystems. But if required, we will do some measurements. And the most contaminated material? We will leave it where it is".

Reception phase:
"OK, if we have to do it let's be smart about it, we minimise dredging and we define several classes of pollution. For the most polluted part of it: just dig a hole or create storage basins such as in the Great Lakes and in Rotterdam" (Figure 9.11).

Constructive phase:
"Recycling and re-use of dredged sediment for brick making or whatever and separating the contaminants from the sediments" (Figure 9.12). Thus new products are developed.

Vision phase:
"Reduce at the source and make the producers of the contamination liable, let them pay for your problems, they are their problems, join the environmental movement in the London Convention". In this phase new coalitions develop, such as a coalition of Port Authorities and Greenpeace and Friends of the Earth. They are your friends when it comes to reduction of polluted dredged material.

Moving from "end-of-pipe" and efficiency measures (from 1960 onwards) to green products and systems innovation (from 1990 onwards) reflects a societal development from reactive to proactive environmental policies.

A transformation to more sustainable systems is only partially a matter of technology. Economic, socio-cultural and institutional changes play an equally important role.

Figure 9.11. The Slufter, a confined storage basin for dredged materials near Rotterdam.

Figure 9.12. Synthetic bricks made from dredged materials, Hamburg, Germany.

Transformation can only be successful when societal and technological changes are mutually reinforcing at different levels, as illustrated in Figure 9.13. This includes the micro scale (niches), the meso scale (regimes) and the macro scale (landscapes) (see Kemp et al., 2000).

Industrial Transformation, a system change, is usually initiated as the result of a local or national innovation, serving as a technological and/or institutional "niche market". When an innovation fits into a regime change that occurs at a regional or continental scale, the

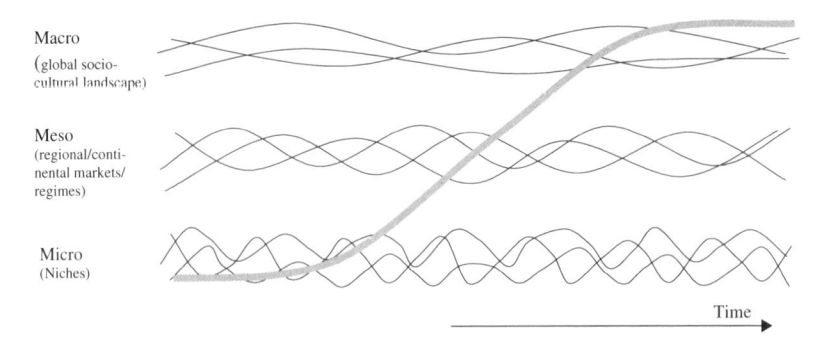

Figure 9.13. Industrial Transformation occurs through mutually reinforcing technological and societal changes at micro, meso and macro scales (Kemp et al., 2000).

innovation is reinforced. When, at the international level, socio-cultural changes occur that favour a new way of behaviour, the system innovation can be absorbed at the global level.

9.9 IMPLICATIONS FOR THE DREDGING SECTOR

Going from the more generic issues about environment and development, we now examine what this implies for the dredging sector.

Coevolution
Coevolution is attractive from the perspective of continued activity of the sector, but this requires a deeper understanding of the ecological processes: how to interact with nature instead of how to beat nature should be the slogan. The WWF plans for the extension of the Rotterdam Maasvlakte are a beautiful example of what can be achieved.

This requires the introduction of ecological expertise in the traditional engineering companies towards eco-engineering. Eco-engineering as much as possible in line with natural systems can help to find solutions supported by more people and decision-makers. Moreover, such projects and schemes are often scientifically and engineering-wise more challenging than many of the traditional engineering solutions.

Triple P performance
Socially more responsive behaviour of companies will help to create trust and thus work for companies. Triple P performance with a balanced focus on Profit, Planet and People could help to make companies more attractive for employees, for shareholders and for clients.

Sustainability concerns
Strategic anticipation of future markets as triggered by sustainability concerns. The major environmental concerns relevant for the dredging industry are:

a. Climate change: how to deal with changing rainfall and run-off patterns requiring adjustments in river, canal, sewage and dike systems and how to deal with changing sea level and storm-surge conditions in estuaries and open coasts; how to contribute to a transition of the energy system: wind power at sea; new gas fields at deeper water;

Figure 9.14. The challenge for the dredging industry is to operate in the spirit of Coevolution with benefits for transportation, shipping, recreation and the natural environment.

CO2-underground storage; new energy networks (piping systems); new harbours for the shipment of biomass and possibly LNG and in the longer future H2.
b. Loss of natural habitats in particular in the coastal zone: how to initiate and develop schemes of combined land use changes and reclamation in the spirit of Coevolution with benefits for the transport sector, for urban development, for industrial siting on the one hand and benefits for recreation for natural habitats and for bioremediation on the other.
c. Water: what can the sector contribute to the solution of the multitude of water related problems now and in the future? Water recharge and water storage will become increasingly important as demand grows, scarcity increases and the climate becomes less predictable. Water transport systems are likely to become more important as well as seawater desalination facilities.
d. Pollution and polluted soils and dredging material: are important now, but major changes in the ways to deal with dredged sediment may be ahead. Reduction at source has gained ground; the emphasis is changing from chemicals concentration as a criterion to biological effects as a criterion for controlled dredging, dumping and storing; a river basin and adjacent seas integrated analysis of environmental effects may well change the predominant policy views on how to deal with the sediments accumulating in harbour basins and shipping channels.

9.10 CONCLUSION

Many companies may already be aware of some of the information presented here. However, how many companies have experts on staff who have an understanding of biology and ecology of intertidal systems, mangrove forests, coral reefs, sea grass systems?

The challenge for dredging professionals is not only the project but also how to meet the needs of people in the field of shelter, transport, water, natural habitats, whilst enhancing the quality of natural systems and biological diversity. The real challenge is to contribute to society whilst ensuring profit and continuation for the firm (Figure 9.14).

This implies that:

- Adhering to the concept of social-natural systems Coevolution with ensured benefits, both for the human species and ecosystems and natural habitats in design, operation and staffing;
- Triple P performance as the bottom line for the company: Profit, Planet and People as a way to create legitimacy for growth amongst clients, shareholders, employees and society at large.

Hopefully these ideas on an overall conceptual approach and on some foreseeable priorities and changes relevant for the dredging sector will stimulate thoughts and innovation capabilities.

Index

A
Accuracy of dredging 8–9, 11–12, 14–17, 20, 22–23, 27–28
Acoustic doppler sediment profiling 45
Aerial photographs 46
Airlift pumping system 178
Amenity 118–119
Archeology 119
Artificial beach 137
Artificial island 196
Auger 96
Auger dredger/dredge 96, 152

B
Backhoe bucket 151
Backhoe dredgers 13, 21, 27, 30, 67, 86, 89, 90–91, 94, 100, 102, 110–111
Barge transport 30, 32, 34
Barges 8, 79, 85, 87, 102, 107, 111, 116, 136, 151, 160, 178, 183, 188, 195, 221
Barge-mounted grab 177
Beach nourishment (replenishment, recharge) 60–62, 64, 115–116, 128–129, 135, 147, 181, 202, 204
Beach recharge, see Beach nourishment
Beach replenishment, see Beach nourishment
Beneficial use (of dredged material) 160, 162
Benthos diversity and biomass 66
Boosters 82
Bottom disc cutter 97
Bottom disc cutter head 96
Bottom disturbance 70
Bow nozzle 177
Breaching 80–81
Bucket ladder dredgers/dredge 12, 27, 30, 84–86, 99, 102, 106, 110, 151, 175, 177, 183, 188, 195

C
Cables 119
Cable burying 59, 62
Cable closing 101–102
Capital dredging 2, 105–106, 108–110, 112, 115, 127–128
Capping, capping techniques 36, 155–157, 189, 191
Chain ladder dredge 151

Clamshell dredge 151
Clay fill 205
Coastal protection 140
Coevolution 225, 230
Combined dredging techniques 148
Combined transport cycles 33
Confined disposal 161–162
Conservation 118–119
Contained aquatic disposal 156–157
Containment (of polluted sediment) 154
Contaminants 18
Contaminated sediment 83, 91, 93, 103, 153–156
Contamination 18, 24, 29–30, 33, 35–36, 70, 102, 119–120, 134–135, 153–154, 170, 172, 183, 188–192, 220
Controlled dumping 136
Conveyor belt transport 32, 188
Cumulative overflow loss 74
Cutter head dredging 148
Cutter suction dredger/dredge 7, 9, 61–63, 67, 74–77, 96, 110, 150–151, 175, 177–178, 181, 183, 195, 213
Cutter suction pipeline dredge 160–161
Cutting modes 76

D
Debris 94–95
Degassing 19, 22–23
Degassing installation 94–95
Density flow 81–83
Deposition of sediment 83
Dike construction 62
Dipper dredge 151
Disc bottom cutter dredger 22
Dilution 8–9, 11, 13–17, 22, 24, 26–28
Dimensionless hopper load 73
Dimensionless overflow rate 73–74
Dispersion (sediment, pollutants) 172
Disposable dredges material (beneficial use of) 121
Disposal 109–115, 119
Dragline bucket 151
Dredge size 152
Dredge spoils 228

Dredgers/dredges for contaminated sediment 154
Dredger/dredge building industry 106
Dredges/dredgers (total number of) 151
Dredges (US) 151–153
Dredging companies (US) 150–151
Dredging costs 145–146, 148–149
Dredging plant 188, 194–195, 222
Dredging plume (see also: Overflow plume. Suspended sediment plume) 185–186
Dredging volumes 64, 109–117, 126, 127–129, 134–137, 147, 149–150, 167, 172–175, 177, 191, 195, 204–205, 207–215, 221
Dump 70
Dumping 130
Dust-pan dredger 178

E
Echo sounders 46
Eco-engineering 238
Eelgrass 42, 44, 46–49
Environmental auger dredger 24
Environmental considerations 82
Environmental effects: criteria 4
Environmental grab dredger 26
Environmental impact assessment/statement/review 39–40, 43, 49, 131, 143, 171, 183–184, 187–189, 191
Environmental investigation, survey 42
Environmental requirements 42
Environmental survey 184–185, 220
Environmental variables 43–44
Environmental windows 143–144
Evaluation of effects 130–131
Excavation 5
Exploration 176–180, 188
Explosives 194

F
Feedback monitoring 43
Feeding (beaches) 116
Filling 172, 189, 185
Fishing 118–119, 221
Floating silt screen 136
Floating pipeline 8, 71, 117, 177
Flow number 78–79

G
Geological research, see Exploration
Geotextil curtains 136
Grab dredgers 14, 21, 27, 30, 67, 87, 88, 94, 100, 102, 110–111, 175, 183, 188, 193, 195
Grab hopper dredger 107
Grab sampling 185

Gravel extraction 60, 64, 65
Groundwater quality 7
Guidelines, see Legislation

H
Hind cast modelling 46, 48
Hopper barge 212
Hopper dredging/dredgers/dredge 150, 148, 160
Hopper transport 30, 32, 34–35
Horizontal positioning 103
Horizontal transport 6
Hydraulic backhoe dredger/dredge 136
Hydraulic closing 101–102
Hydraulic dredgers/dredge 7, 19, 151
Hydraulic pipelines 81
Hydrodynamic dredgers 15, 66

I
Illegal disposal 220
Impeller eye 96
In-situ capping 155, 157
Integrated Coastal Zone Management 118
Intertidal flats 3

J
Jet pumps 71

L
Land reclamation (see also Reclamation) 76
Landfill 42
Lean mixture overboard system 191
Legislation (licences, regulations, guidelines) 108, 111–113, 115, 117–120, 129–133, 170–172, 186, 221
Level bottom capping 155–157
Licences, see Legislation
Light attenuation 45
Loading 71–72
Loose spill layers 8–9, 11, 13–17, 22, 24–25, 27, 29
Loss of coral reef 218–219

M
Maintenance dredging 3, 18, 60–61, 64, 106–108, 111, 113–116, 127–128, 148, 174–175, 202, 204
Management 176, 178, 191
Marine aggregates 117
Marine environment 118–119
Mechanical dredgers/dredges/dredging 12, 20, 148, 151, 160
Mechanical excavation 66
Mitigation measures 36, 37
Mixing of soil layers 8–9, 11–12, 14–17, 22, 24–25, 27, 29

Mobile dredging plant 177
Modeling dredging, disposal 163–164, 172, 220
Monitoring 18–20, 38–45, 102, 109, 171, 177, 185–186, 193, 220
Monitoring strategy 43, 45
Mud dispersal 182, 186–191, 193, 195
Mud dredging 174, 181, 186, 188, 191, 195
Mudflats 219
Mussels 44–45, 48–49

N
Navigation 118–119
Noise and air pollution 6–8, 10–11, 13, 23–24, 26–29, 32, 177, 220

O
Open water disposal 160–161
Optical backscatter 45
Output rate 8–11, 13–17, 23–24, 26–27, 29
Overflow 6, 19, 87, 96, 117, 177, 183, 191
Overflow loss 72–75
Overflow plume 177, 184–185
Overflow procedure 160

P
Phases of dredging 5–7
Pipeline 79, 81, 96, 116, 119, 135, 181, 202, 217, 218
Pipeline burying 59, 62
Pipeline transport 29–32, 35
Placement, placing dredged material 6–7, 147, 220
Placement on land 33
Plain suction dredger/dredge 61, 63, 67, 75, 79, 80–83, 151
Pneumatic dredger 28–29
Polluted sediment depot 135–136
Port development 61, 202
Port dredging 127, 157
Port extension 61
Positioning 102–103

R
Rainbowing 208
Reclamation 45, 64, 70–71, 128, 130, 172, 174–177, 181–184, 186–188, 194–196, 201–215, 218–222
Re-colonisation 109
Relative cumulative overflow loss 72–73
Remedial dredging 3–4, 19, 26, 29
Replenishment 115–116
Research (on dredging) (US) 164
Reserve calculations 179–180
Re-suspension of sediment 83, 91, 114

Road transport 31
Rubber cover 101

S
Safety 8–9, 11–12, 14, 16–17, 22, 24–25, 27–28, 221
Sand dredging 172, 175–177, 178, 181, 183, 186
Sand extraction 63–66
Sand fill 213
Sand grain angularity 180
Sand mining 202, 204–205
Sand pump dredgers 106
Satellite photographs 46
Scow 151, 160
Scraper head 96, 98–99
Seagrass beds 219
Sediment classification 171
Sediment dispersion 43
Sediment disposal 205
Sediment plumes 111, 117
Sediment specifications 183
Sediment storage 61
Sediment supply 114
Sediment traps 45–46
Sedimentation 43, 45
Sedimentation rate 218–219
Seismic survey 178–179
Shell extraction 59, 65
Shingle recharge 116
Shore connection 71
Shore protection 61
Shoreline changes 221
Shovel type dredgers 94
Sidecasting dredge 151
Sidescan sonar 46–47, 66
Silt screens 103, 220
Siltation 221
Siltation rate 113
Slice head 96, 98–99
Smell 220
Spatial (coastal) planning 118
Spillage 31, 78
Spray pontoons 213
Static dredging plant 177
Stationary cutter suction dredger 75
Stationary suction dredgers 7–9
Storage of sediment 61
Storage of polluted sediment 61
Study of management and location 131
Submarine video 46
Suction dredgers 106, 110
Suction pipe 98
Suction tube 79–80
Survey 111

Suspended sediment, Suspended matter, turbidity 3–5, 8–9, 11–12, 14–17, 19, 22, 24–25, 27–28, 34–35, 45–46, 72, 83–84, 88, 90, 96–97, 99–100, 102, 111–113, 132–133, 172, 177, 184–187, 191, 193, 218, 221
Suspended sediment plume 184
Sweep dredger 23
Sweep head 96, 98
Swing direction 97
Swing pattern 76

T
Territorial waters 222
Trailing/Trailer suction hopper dredger 10, 19, 23, 30, 34, 61, 62, 64, 66–69, 76, 93–94, 96, 107, 111, 117, 135, 175, 177–178, 181, 183, 188, 193–195, 212
Transport by train 206

Truck transport 151
Turbidity, see Suspended sediment
Turtle deflection device 19

U
Underwater placement 34, 36
Underwater plough 17–18
Unloading 31

V
Vibration 220

W
Water injection dredgers/dredge 16, 83, 107, 151
Water jets 81
Water quality 42–44
Well covered bucket 101
Wetlands 3